ELECTRICAL MATERIALS

SECOND EDITION

ELECTRICAL MATERIALS

SECOND EDITION

Rob Zachariason

DELMAR
CENGAGE Learning·

Australia • Brazil • Japan • Korea • Mexico • Singapore • Spain • United Kingdom • United States

Electrical Materials, 2E
Rob Zachariason

Vice President, Editorial: Dave Garza

Director of Learning Solutions: Sandy Clark

Acquisitions Editor: Stacy Masucci

Managing Editor: Larry Main

Senior Product Manager: John Fisher

Editorial Assistant: Andrea Timpano

Vice President, Marketing: Jennifer Baker

Marketing Director: Deborah Yarnell

Marketing Manager: Katie Hall

Marketing Coordinator: Jillian Borden

Senior Production Director: Wendy Troeger

Production Manager: Mark Bernard

Senior Content Project Manager:
    Jennifer Hanley

Senior Art Director: David Arsenault

COVER CREDITS:

BACKGROUND IMAGE:

© Jim Jurica/iStockphoto.com

INSET IMAGES: (clockwise, from top left)

© Cutcaster.com/P?

© Georg Hanf/iStockphoto.com

© Cengage Learning 2013

© VIPDesignUSA/Shutterstock.com

© Albln/Dreamstime.com

© REDAV/Shutterstock.com

For product information and technology assistance, contact us at
Cengage Learning Customer & Sales Support, 1-800-354-9706
For permission to use material from this text or product,
submit all requests online at **www.cengage.com/permissions**.
Further permissions questions can be e-mailed to
permissionrequest@cengage.com

Library of Congress Control Number: 2011930762

ISBN-13: 978-1-111-64006-4

ISBN-10: 1-111-64006-8

Delmar
5 Maxwell Drive
Clifton Park, NY 12065-2919
USA

Cengage Learning is a leading provider of customized learning solutions with office locations around the globe, including Singapore, the United Kingdom, Australia, Mexico, Brazil, and Japan. Locate your local office at:
international.cengage.com/region

Cengage Learning products are represented in Canada by Nelson Education, Ltd.

To learn more about Delmar, visit **www.cengage.com/delmar**

Purchase any of our products at your local college store or at our preferred online store **www.cengagebrain.com**

Notice to the Reader

Publisher does not warrant or guarantee any of the products described herein or perform any independent analysis in connection with any of the product information contained herein. Publisher does not assume, and expressly disclaims, any obligation to obtain and include information other than that provided to it by the manufacturer. The reader is expressly warned to consider and adopt all safety precautions that might be indicated by the activities described herein and to avoid all potential hazards. By following the instructions contained herein, the reader willingly assumes all risks in connection with such instructions. The publisher makes no representations or warranties of any kind, including but not limited to, the warranties of fitness for particular purpose or merchantability, nor are any such representations implied with respect to the material set forth herein, and the publisher takes no responsibility with respect to such material. The publisher shall not be liable for any special, consequential, or exemplary damages resulting, in whole or part, from the readers' use of, or reliance upon, this material.

Printed in the United States of America
1 2 3 4 5 6 7 15 14 13 12 11

I would like to thank my wife, Brandi, and my children Lauren, Kate, and Julia, for the understanding and support they gave me as I worked on revising this text.

Thank you to all the people at Delmar Cengage Learning for their hard work, but in particular Stacy Masucci and John Fisher.

CONTENTS

INTRODUCTION

The primary purpose of this book is to help people new to the electrical industry identify the industry's commonly used materials and understand its terminology. Learning how to navigate and apply the *National Electrical Code®* is a difficult task in itself, even for someone who is familiar with the materials and terms. Thus, this text provides pictures of the materials and brief descriptions of them.

The book is in outline format and contains color photos, making the materials easy to visualize and the references quick and to the point. Unfortunately, this book cannot list every part, or every style of a part, that electricians encounter. So, I have included the parts that are commonly used, to provide readers with at least a foundation of knowledge. Once readers understand the basic types of materials, additional information can be more easily understood.

Many of the parts used in the electrical industry have different names in different areas of the country. Sometimes the name of a part may be the manufacturer's name; at other times, there does not seem to be any rhyme or reason to why a part received its name. This is referred to as "trade slang," and I have included some of these terms in parentheses.

Some of the materials noted in this book are accompanied by a code reference. This reference indicates the article in the *National Electrical Code®* that applies to the product as a general rule. *Be aware that this may not be the only code article in the NEC® that applies to that particular item.* The code references in this text are based on the *2011 National Electrical Code®*.

ABOUT THE AUTHOR

Rob Zachariason is a full-time electrical instructor at Minnesota State Community and Technical College. He is a member of the International Brotherhood of Electrical Workers, the International Association of Electrical Inspectors, and the National Education Association. He also teaches an electrical apprenticeship course for the Joint Apprenticeship and Training Committee.

Rob is a graduate of the Construction Electricity program at the former Northwest Technical College, as well as of the Inside Wireman program for the National Joint Apprenticeship and Training Committee. He worked as a journeyman electrician, a service electrician, and a foreman for 10 years before becoming a full-time instructor.

ACKNOWLEDGMENTS

The author and Delmar Cengage Learning wish to acknowledge and thank the members of our review panel for their suggestions during development of this book. Thanks go to:

Kevin Boiter
Piedmont Technical College
Greenwood, South Carolina

Glenn Gibert
Isothermal Community College
Spindale, North Carolina

William Quimby
Shelton State Community College
Tuscaloosa, Alabama

Richard Schell
Luzerne County Community College
Nanticoke, Pennsylvania

John Schroader
Owensboro Community College
Owensboro, KY

CHAPTER

1

CONDUCTORS AND CABLES

INTRODUCTION

Conductors and cables are constantly changing as new technology allows for new and improved conductor insulation and cable sheaths. This chapter gives a brief introduction to and covers some of the commonly used conductors and cables.

Conductor versus Cable

What is the difference between a conductor and a cable? A conductor is a conductive material that allows electric current flow. This chapter will look at conductors in the wire form; however any object that allows the flow of electricity could be a conductor, for example bus bars, raceways, metal frames, and so on. A conductor may or may not have insulation. The majority of the conductors used in the electrical industry for branch circuits and feeders are insulated.

A cable is a factory assembly of conductors. Most cables have the conductors protected by a sheath; however, this is not always the case. Some cables are conductors that are simply twisted slightly together with no sheath, like underground service-entrance or service-drop cable.

Conductor Insulation and Cable Sheath

Conductor insulation and cable sheath are not the same thing. Conductor insulation encapsulates a wire to prevent it from contacting other conductors or objects. A cable sheath surrounds a factory assembly of conductors to offer protection for the conductors and keep them bundled together.

There are various types of conductor insulation available to meet the different conditions, environments, and voltages that conductors will be exposed to (Figure 1-1). However, although there are many types of conductor insulation available, there are only a few that are commonly used. The others are used only when a specific need arises or are a type of insulation that was used in the past. *NEC® Table 310.104* gives specific information about most of the available conductor insulations.

Cables will often have a sheath to meet the conditions, environments, and physical abuse they will be subjected to. Appropriately rated cables may be installed within building structures, underground, overhead, and so on, without the use of a raceway for protection. Individual conductors (building wire) require the use of a raceway to house and protect the conductors.

R	Rubber	No "H"	60 degrees Celcius
T	Thermoplastic	One "H"	75 degrees Celcius
X	Cross linked polyethelyne	Two "HH"	90 degrees Celcius
S	Silicone		
W	Wet Location	−2	90 degrees wet or dry location
N	Nylon outer covering		

Examples	
THHN	Thermoplastic insulation with a nylon covering rated for 90 degrees Celcius
RHW	Rubber insulation rated for wet lcoations and 75 degrees Celcius

© Cengage Learning 2012

FIGURE 1-1

Smaller																				Larger
26	24	22	20	18	16	14	12	10	8	6	4	3	2	1	1/0	2/0	3/0	4/0	250 kcmil	300 kcmil	350 kcmil	400 kcmil	500 kcmil	etc.
American Wire Gauge (AWG)																			Circular Mils					

© Cengage Learning 2012

FIGURE 1-2

Sizes

Conductors are available in different sizes and will have a maximum current-carrying capacity. The methods used to size conductors are the American Wire Gauge (AWG) and circular mils. The American Wire Gauge is set up so that the larger the number, the smaller the wire. For example, a #14 gauge conductor is smaller than a #2 gauge conductor. The American Wire Gauge is only used to size conductors up to 4/0 (four ought); above that, conductors are sized by their area in circular mils, starting with 250,000 circular mils (kcmils) (Figure 1-2). Although smaller conductors are sized in AWG, they also have a circular mil area, which is detailed in *Chapter 9, Table 8*, of the *National Electrical Code®*. Each size conductor will have a maximum current it is capable of carrying without its insulation overheating. This will be determined by many factors, such as insulation type, ambient temp, number of conductors grouped together, and whether they are located in a wet, damp, or dry location.

Solid versus Stranded

Conductors are available solid or stranded. The *NEC®* requires conductors #8 and larger that are installed in a raceway to be stranded. Conductors #10 and smaller are typically available in both solid and stranded. Solid conductors have an advantage of being terminated directly to most devices, whereas stranded conductors are easier to pull through raceways and can be crimped into terminals. Finely stranded conductors are used in cords and where flexibility is necessary, as it will allow for movement without loosening connections. Some specifications will prescribe solid or stranded conductors; otherwise, it simply comes down to personal preference.

Compact conductors are stranded conductors that have squared off rather than round strands (Figure 1-3). Having squared-off strands allows the air space that is found between round strands to be eliminated, thereby reducing the total area of the conductor and making it smaller. Not all conductors are available this way, only larger conductors in specific types of insulation. *Table 5A* in *Chapter 9* of the *National Electrical Code®* specifies the area of compact conductors.

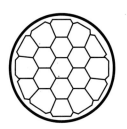

Compact strand conductor

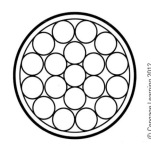

Circular strand conductor

© Cengage Learning 2012

FIGURE 1-3

Copper versus Aluminum

Conductors are available in copper, aluminum, and copper-clad aluminum, with copper and aluminum being the most common. Each type of conductor material has its own advantages and disadvantages.

Copper conductors have the advantage of having a lower resistance and better terminating characteristics than aluminum conductors; however, they are more expensive and heavier. Having a lower resistance means that a smaller size conductor can be used to pass the current without creating excess heat. For example, 50 amps of current would require #8 copper or #6 aluminum conductor.

Aluminum conductors have the advantage of costing less and weighing less; however, they have a higher resistance than copper and require special care when terminating. Aluminum oxidizes when exposed to the air, so an anti-oxidizing paste must be used on all terminations. With a change in temperature, aluminum also expands and contracts more than copper, which can cause loose connections if they are not terminated properly.

It is not permitted to mix aluminum and copper together as a galvanic reaction will occur and ultimately lead to a loose connection. The exception to this rule is listed connectors that are rated to mix aluminum and copper.

Building Wire

| #12 THHN Solid Copper | #4 Bare Stranded Copper | 2/0 XHHW Stranded Aluminum | 750 KCmil THW Stranded Aluminum |

© Cengage Learning 2012

1. Building wire is available in solid or stranded.
2. Building wire is available in copper or aluminum.
3. Building wire must be installed in a raceway (with the exception of a few grounding situations).
4. Building wire is available in many different types of insulation to handle all the various needs.
5. Building wire with the most common insulation used for general wiring is dual-rated THHN/THWN.
6. Building wire is available with many colors of insulation:

   a. black
   b. red
   c. blue
   d. white
   e. brown
   f. orange

   g. yellow
   h. gray
   i. pink
   j. orange
   k. purple
   l. green

7. Bare conductors are used for grounding and bonding.
8. Some types of building wire are available with compact strands, which makes the overall conductor diameter smaller.

   Note that insulated conductors larger than #10 are typically black and are marked with colored phase tape according to their phase or purpose. Some manufacturers have the larger conductors available with colored insulation by special order.

NEC® Article 310, Conductors for General Wiring
NEC® Article 250, Grounding

Nonmetallic Sheathed (NM) Cable (Romex®)

14-2   14-2-2   12-2   10-3

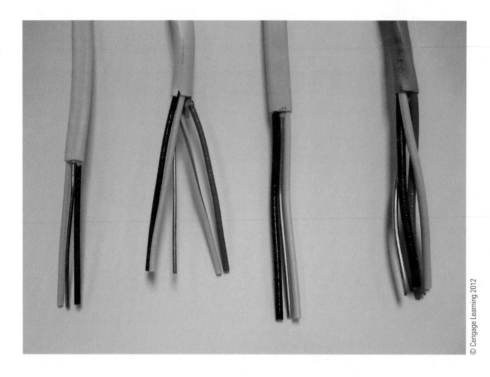

© Cengage Learning 2012

Definition: A factory assembly of two or more insulated conductors enclosed within an overall nonmetallic jacket. *NEC® 334.2*

1. NM cable is typically used in residential and small commercial applications.
2. All new NM cable comes with a grounding conductor (bare copper). This bare conductor is not included when counting the number of conductors in the cable. For example, the "14" in 14-2 NM indicates that there are 14-gauge conductors in the cable, and the "2" indicates that there are two insulated conductors in the cable.
3. NM cable comes in color-coded jackets to make identification of wire size easier:
    a. white   14 gauge
    b. yellow   12 gauge
    c. orange   10 gauge
    d. black   8 and 6 gauge
4. NM cable comes in different wire configurations:
    a. 2 wire with ground   white, black, bare copper
    b. 3 wire with ground   white, black, red, bare copper
    c. 4 wire with ground   white, black, red, blue, bare copper
    d. 14-2-2 wire with ground   black, white, red, white/red stripe, bare copper
5. NM-B cables have conductors rated for 90° C; however, the 60° column is used when finding ampacities.

    Note that current-carrying conductors 14-10 gauge are solid and that current-carrying conductors 8 and larger are stranded.

NEC® Article 334, Nonmetallic Sheathed Cable

Underground Feeder (UF) Cable

12-2 UF   10-3 UF

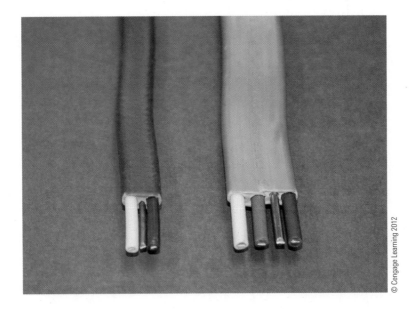

Definition: A factory assembly of one or more insulated conductors with an integral or an overall covering of nonmetallic material suitable for direct burial in the earth. *NEC® 340.2*

1. UF cable typically is used in residential applications to run power out to a small load, such as a post light, river pump, and so on.
2. UF cable has conductors molded directly into the nonmetallic jacket.
3. UF cable is rated to be buried directly in the earth, with no raceway.
4. UF cable is resistant to sunlight.
5. UF cable has wire configurations similar to nonmetallic sheathed cable:
    a. 2 wire with ground   white, black, bare copper
    b. 3 wire with ground   white, black, red, bare copper

Note that current-carrying conductors 14-10 gauge are solid and that current-carrying conductors 8 and larger are stranded.

Also, newer cables have conductors rated for 90° C; however, the 60° column is used when finding ampacities.

NEC® Article 340, Underground Feeder and Branch Circuit Cable

Thermoplastic Cable

18-2   18-5   18-8 FPLP

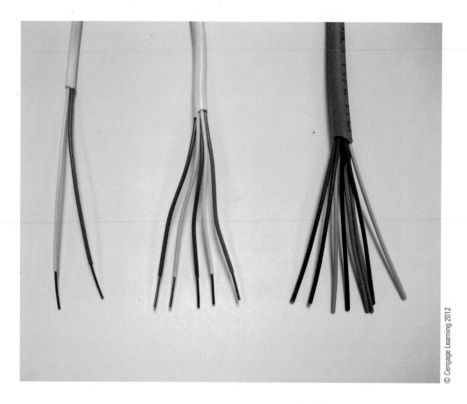

© Cengage Learning 2012

1. Thermoplastic cable is typically used for control wiring:
   a. heating and cooling controls
   b. chimes
   c. garage door openers
   d. fire alarm controls (if rated appropriately)
2. Thermoplastic cable is used for applications 50V and under.
3. Thermoplastic cable does not have a grounding conductor like NM cable. The first number indicates the conductor size, and the second number indicates the number of conductors. If there is an initial after the size and number of conductors, it indicates what rating the jacket has.
   a. general purpose   X
   b. plenum rated   P
   c. riser rated   R
   d. fire alarm   F
   e. tray cable   TC
   f. non-power limited   NPL
   g. power limited   PL

   Example: FPLP—fire alarm, power-limited, plenum-rated cable
4. Thermoplastic cable is available with solid or stranded conductors.

NEC® Article 725, Class 1, Class 2, and Class 3 Remote-Control, Signaling, and Power-Limited Circuits
NEC® Article 760, Fire Alarm Systems

Armored Cable (AC) (BX)

12-3 AC Cable

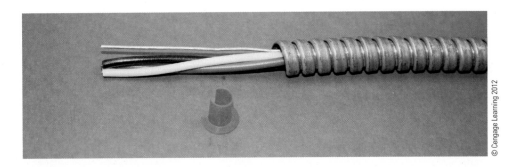

© Cengage Learning 2012

Definition: A fabricated assembly of insulated conductors in a flexible interlocked metallic armor. *NEC® 320.2*

1. AC has a bare aluminum conductor for bonding.
    a. Bonding conductor is smaller gauge than the current-carrying conductors.
    b. Armored cable must have fittings listed to provide a grounding path.
2. AC is typically used for general wiring in commercial applications.
3. AC comes in many wire configurations and can be special ordered with special size and color combinations.
4. AC has the following typical configurations:
    a. 2 wire   black, white, bare aluminum bonding conductor
    b. 3 wire   black, red, white, bare aluminum bonding conductor
5. AC is available with copper and aluminum current-carrying conductors.
6. AC is labeled similarly to NM cable: "12-3" indicates there are three 12-gauge current-carrying conductors in the cable. There is also a bare aluminum bonding conductor.
7. AC is available with a steel or aluminum sheath.
8. A red bushing is installed where the conductors leave the metal sheath to prevent abrasion.

Note that newer cables have conductors rated for 90° C; however, the 60° column is used when finding ampacities.

NEC® Article 320, Armored Cable

Metal-Clad (MC) Cable

12-2 MC

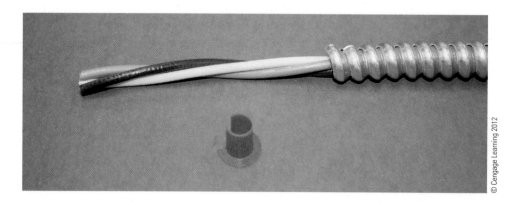

© Cengage Learning 2012

Definition: A factory assembly of one or more insulated circuit conductors with or without optical fiber members enclosed in an armor of interlocking metal tape, or a smooth or corrugated metal sheath. *NEC® 330.2*

1. MC cable has an insulated green grounding conductor.
2. MC cable is typically used in commercial applications.
3. MC cable comes in many wire configurations and can be special ordered with special size and color combinations.
4. MC cable has these typical configurations:
   a. 2 wire   black, white, green grounding conductor
   b. 3 wire   black, red, white, green grounding conductor
5. MC cable is available with two green grounding conductors for isolated ground systems, in which case one of the green conductors has a yellow stripe.
6. MC cable typically has copper conductors.
7. MC cable is labeled similarly to NM cable: "12-2" indicates that there are two 12-gauge current-carrying conductors in the cable. There is also a green grounding conductor.
8. MC cable is available with a PVC coating that may allow it to be used in the following locations:
   a. direct burial
   b. wet locations
   c. concrete embedment
   d. direct sunlight
9. A red bushing is installed where the conductors leave the metal sheath to prevent abrasion.

   Note that newer cables have conductors rated for 90° C.

NEC® Article 330, Metal-Clad Cable

Service-Entrance (SE) Cable

SER   SEU

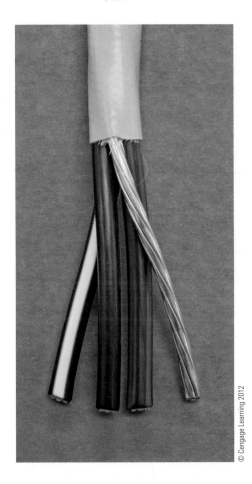
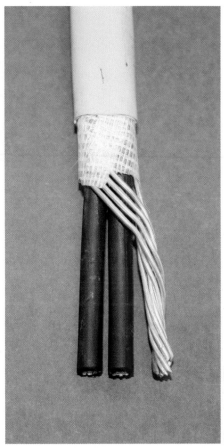

Definition: A multiconductor assembly provided with an overall covering having a flame-retardant, moisture-resistant covering that is primarily used for services.

1. SE cable typically is used in residential applications:
   a.   feeds from the meter socket to the main panel
   b.   feed from the main panel to a subpanel
   c.   feed large appliances, such as a range, plenum heater, and so on
2. SE cable can be run inside or outside without a raceway.
3. SE cable is available in aluminum and copper.
4. SE cable with three insulated conductors and a bare grounding conductor is type R (SER).
5. SE cable with two insulated conductors and a bare grounding conductor is type U (SEU).
6. SE cable with either a grounding conductor that is formed into a single conductor, as in the SER above, or braided around the current-carrying conductors, as in the SEU above.

NEC® Article 338, Service-Entrance Cable

Underground Service-Entrance Cable (USE)
(Underground Utility Cable) (Underground Triplex)
[URD Underground Residential Distribution]

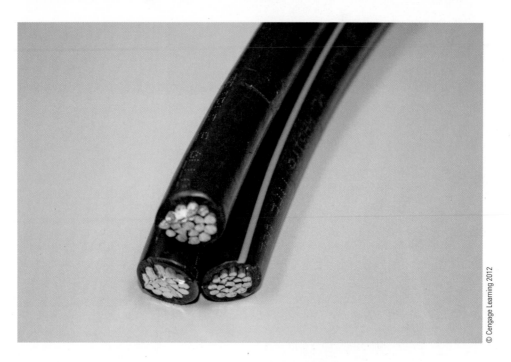

© Cengage Learning 2012

Definition: A moisture-resistant single conductor or multiconductor assembly listed for underground use that is primarily used for services.

1. Underground service-entrance cable (USE) is rated to be directly buried without a raceway.
2. USE typically is used by the power company to bring power from the utility transformer to a residence.
3. USE is used by electricians to bring power to detached buildings or from meter pedestals to buildings.
4. Larger cables are available with a full size or derated neutral conductor (conductor with the yellow stripe).

NEC® Article 338, Service-Entrance Cable

Service-Drop Cable (Overhead Triplex)

© Cengage Learning 2012

1. Service-drop cable is used to provide overhead power from a utility pole to a building.
2. Service-drop cable may have two or three insulated conductors and one bare conductor.
3. Service-drop cable has a steel messenger wire in the bare conductor that can withstand the tension of running overhead.

   WARNING: Do not cut the bare messenger wire with regular cable cutter; doing so will notch the cutting edge.

NEC® Article 396, Messenger-Supported Wiring

Mineral-Insulated (MI) Cable

MI Cable with HDPE Jacket

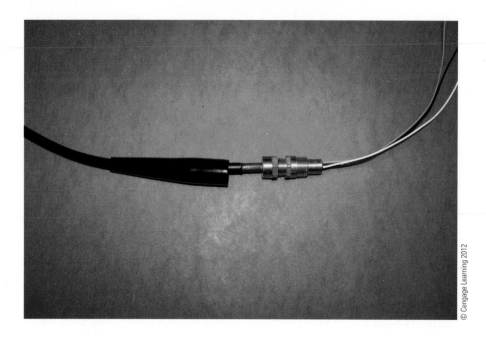

© Cengage Learning 2012

Definition: A factory assembly of one or more conductors insulated with a highly compressed refractory mineral insulation and enclosed in a liquidtight and gastight continuous copper or alloy steel sheath. *NEC® 332.2*

1. MI cable has conductors imbedded in magnesium oxide with a copper or alloy sheath.
2. MI cable is typically used in hazardous locations as it will not allow the passage of vapors, gases, liquids, and flames.
3. MI cable is available with a high-density polyethylene jacket for copper-corrosive environments.
4. MI cable is radiation resistant and waterproof.
5. MI cable without a HDPE jacket is fire and heat resistant.
6. Ends must be properly terminated and sealed to prevent entrance of moisture.

NEC® Article 332, Mineral-Insulated Metal-Sheathed Cable

Flexible Cord (Portable Cord) (Rubber Cord) (SO Cord)

14-8 SOW Cord   12-3 SO Cord

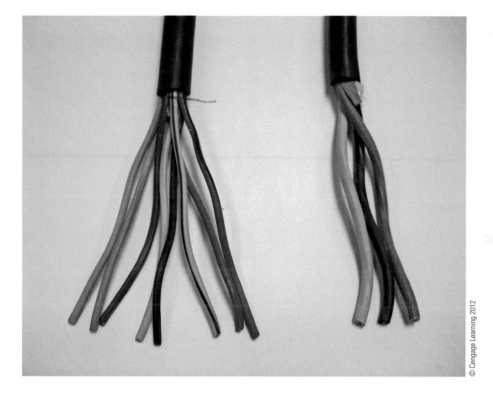

© Cengage Learning 2012

1. Flexible cord is a factory assembly of conductors intended to supply equipment or portable appliances.
2. Flexible cord is used to connect equipment to power supplies where flexibility is needed.
3. Flexible cord comes in many wiring configurations and sizes.
4. Flexible cord is available with different types of jackets (SO, SOW) to withstand different environments and usages.
5. Flexible cord may have a plug installed on one or both ends.
6. Flexible cord has stranded conductors to allow maximum flexibility.
7. Flexible cord has conductors that are *not* counted like those in NM cable or MC. Instead, the green grounding conductor is counted with the current-carrying conductors. For example, the 12-3 SO cord in the picture above has three 12-gauge conductors inside:
   a. 1 black
   b. 1 white
   c. 1 green

NEC® Article 400, Flexible Cords and Cables

Communications Cable

Coaxial RG6	D-Station 4 Conductor	Category 3 4 Pair	Category 5E 4 Pair	Fiber-Optic 6 Fiber

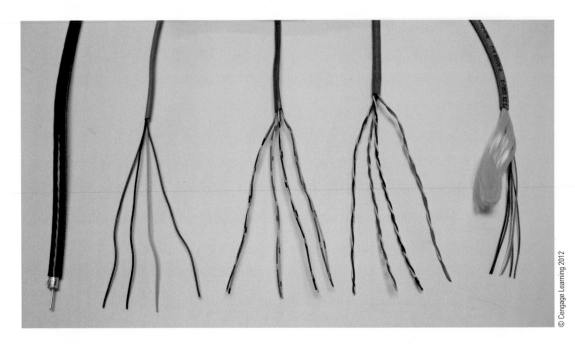

© Cengage Learning 2012

1. RG6 is used to provide TV signals, and in some cases is used as backbone computer cabling.

2. D-Station wire (Bell wire) is rarely used now. In the past, it was used for telephone cabling.

3. Category 3 replaced D-station wire but is already somewhat obsolete. It is still being used, however, in some applications for telephone cabling. This cable typically has four pairs of twisted conductors.

4. Category 5 and Category 5E are used for computer networking as well as for telephone cabling. This cable may have four or five pairs of conductors. Each set of conductors has a different twist length.

5. Fiber-optic cable is used for many aspects of communications because it can carry all the information carried by the other cables listed above and much more on one strand of fiber.

6. Cable markings are as follows:
   a. dwellings   X
   b. general purpose   G
   c. plenum rated   P
   d. riser rated   R
   e. communications   CM
   f. cable television   CATV

   Example: CMP-communication, plenum rated cable

NEC® Article 770, Optical Fiber Cables and Raceways
NEC® Article 800, Communication Circuits
NEC® Article 810, Radio and Television Equipment
NEC® Article 820, Community Antenna, Television, and Radio Distribution Systems
NEC® Article 830, Network-Powered Broadband Communication Systems
NEC® Article 840, Premises-Powered Broadband Communications Systems

Bundled Cable

2 Category 5 Cables
2 Coaxial Cables

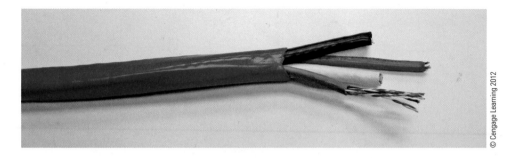

© Cengage Learning 2012

1. Bundled cable is available in many types of cables bundled together:
    a. Cat 5 cable
    b. coax
    c. fiber-optic
    d. Cat 3 cable
2. Bundled cable eliminates having to run each cable separately.

Medium-Voltage Cable

© Cengage Learning 2012

Definition: A single or multiconductor solid dielectric insulated cable rated 2001 volts or higher. *NEC® 328.2*

1. Medium-voltage cable may be rated for voltages from 2,000 to 35,000V.

2. Medium-voltage cable may be rated to be installed in conduit or cable trays, aerial, or directly buried.

NEC® Article 328, Medium-Voltage Cable

Oxide Inhibitor

© Cengage Learning 2012

1. Oxide inhibitors are used when terminating aluminum conductors to prevent oxidation.
2. After the insulation has been removed from an aluminum conductor, it should be covered with an oxide inhibitor as soon as possible, as it will begin to oxidize immediately.

Wire-Pulling Lubricant (Wire Lube)

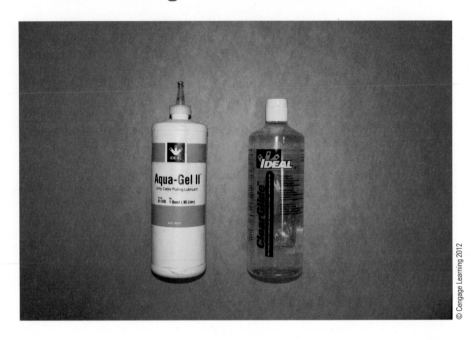

© Cengage Learning 2012

1. Wire-pulling lubricant lubricates conductors so they slip through a raceway more easily.

CHAPTER 2

NONMETALLIC BOXES

INTRODUCTION

Boxes are used as a place to pull wires, splice wires, mount devices, mount lights, and so on. They can be divided into two categories: nonmetallic boxes, and metallic boxes. This chapter will focus on nonmetallic boxes.

Nonmetallic (NM) boxes are typically used with nonmetallic sheathed cable or nonmetallic raceways. There are numerous nonmetallic boxes on the market, but they can be distinguished by the material they are made from, the number of gangs, and the mounting style.

Nonmetallic boxes are available in a variety of materials, with the main three being thermoplastic, phenolic, and fiberglass. Each manufacturer will have slight differences in box design, but they will be essentially the same.

Nonmetallic boxes are available in device (gang) boxes, round boxes, and 4-square boxes. Device boxes are identified by the number of gangs they have. For example: a single-gang box will house one device, a 2-gang box will house two devices, and so on. Device boxes are intended to mount devices, so they have holes designed to fit 6-32 screws. Round boxes are intended to mount luminaires (light fixtures) and have holes designed to fit 8-32 screws. Four-square boxes can be used to mount devices or luminaires; however, the box itself only has screw holes in diagonal corners. A mudring (plaster ring) must be mounted on the box to allow mounting holes for a device or luminaire.

Nonmetallic boxes will have a maximum volume in cubic inches they can house. Items that must be figured into the calculation include conductors, devices, internal clamps, fixture studs, and hickeys. *Article 314.16* of the *NEC*® has detailed instruction on how to perform box fill calculations. Each nonmetallic box will be stamped with its maximum volume (Figure 2-1). As each manufacturer's box varies slightly, the cubic-inch capacity of each box also varies.

It is a requirement of the *NEC*® that cables and raceways be fastened securely to boxes. Nearly all nonmetallic boxes have internal clamps, which must be included in the box fill calculation. The exception to this rule is single-gang boxes, which typically do not have clamps. The *NEC*® allows this as an exception and then requires NM cable to be secured closer to a single-gang box than it would have to be secured with other boxes that do have clamps.

In some parts of the country, local codes require a vapor seal on all boxes mounted in exterior walls and ceilings. Vapor seal boxes have a foam or rubber seal that will be pressed against the polyurethane sheathing in the exterior wall to create the seal. Nearly all manufacturers have a nonmetallic box on

FIGURE 2-1

the market that will satisfy this requirement. Note that it may be necessary to seal around where the conductors enter the box with a sealant that won't deteriorate the cable sheath.

Round boxes that have been designed to mount luminaires will typically have a maximum fixture weight of 50 pounds that they can support. If a very large or heavy fixture is to be installed, a box that allows support independent of the box may be required. A few requirements must also be kept in mind for paddle fans. They shall not be mounted on a regular ceiling box. They must be mounted to a box that is marked "Acceptable for fan support" and has been tested to handle the vibration, torque, and extra weight a paddle fan has.

Device Boxes

| Single-Gang Fiberglass with Vapor Seal | 2-Gang Thermoplastic | 3-Gang Thermoplastic with Vapor Seal |

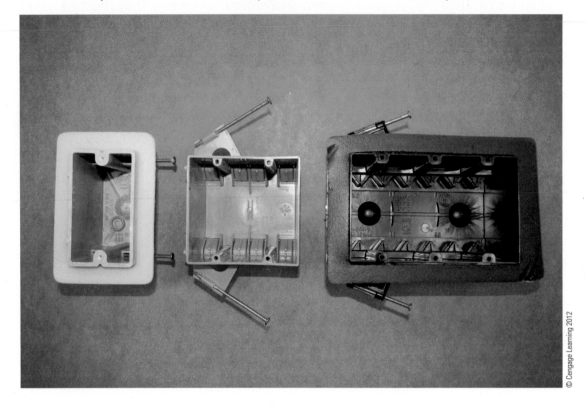

© Cengage Learning 2012

1. Nonmetallic device boxes are used primarily with nonmetallic sheathed cable in residential applications.
2. Nonmetallic device boxes are available in different materials:
   a. fiberglass
   b. phenolic
   c. thermoplastic
3. Nonmetallic device boxes have a specific number of gangs, which determines how many devices can be inserted into the box.
4. Nonmetallic device boxes also come in vapor-sealed boxes, which are required in some parts of the country for all exterior walls and ceilings.
5. Nonmetallic device boxes typically have the cubic-inch capacity stamped inside the box.
6. Nonmetallic device boxes are available in many styles and sizes to fit the needs of various applications.
7. Nonmetallic device boxes are available in 4 gang.

NEC® Article 314, Outlet, Device, Pull, and Junction Boxes; Conduit Bodies; Fittings; and Handhole Enclosures

Screw-in Device Boxes

Single-Gang
Thermoplastic

2-Gang
Thermoplastic

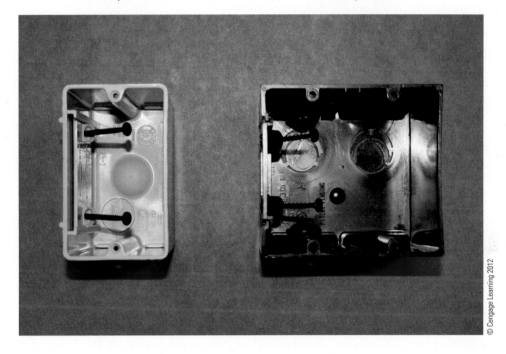

© Cengage Learning 2012

1. Screw-in device boxes can be fastened into place from inside the box.

2. Screw-in device boxes may be used for adding a box into an existing wall if cut in along a stud.

3. The box can be adjusted in or out after the finished wall surface is in place.

NEC® Article 314, Outlet, Device, Pull, and Junction Boxes; Conduit Bodies; Fittings; and Handhole Enclosures

Side Bracket and Pocket Boxes

Side Bracket Thermoplastic   Pocket Box Thermoplastic

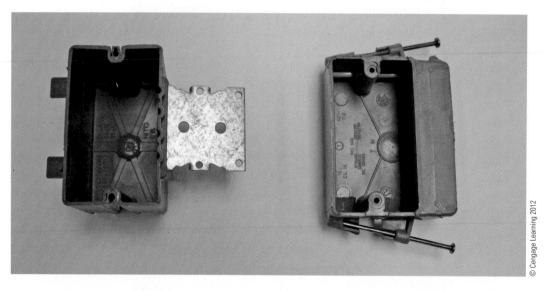

© Cengage Learning 2012

1. Side bracket boxes are used when it is difficult to nail the box in from the side.
2. Pocket boxes are used when there is not enough depth for a single-gang box.
    a. Some pocket boxes have only about a 1/4" rise from the pocket to the front of the box and are designed for paneling.

NEC® Article 314, Outlet, Device, Pull, and Junction Boxes; Conduit Bodies; Fittings; and Handhole Enclosures

Ceiling Boxes

4″ Round Thermoplastic
with Vapor Seal

4″ Round Phenolic

© Cengage Learning 2012

1. Ceiling boxes typically are used for ceiling outlets.
2. Ceiling boxes sometimes come without nails. These can be screwed or nailed on through the flange.
3. Nonmetallic ceiling boxes are available with a 3" or 4" diameter.

NEC® Article 314, Outlet, Device, Pull, and Junction Boxes; Conduit Bodies; Fittings; and Handhole Enclosures

Round Barhanger (Phenolic) Box

© Cengage Learning 2012

1. Round Barhanger boxes are used when a lighting outlet must be placed between two studs.

NEC® Article 314, Outlet, Device, Pull, and Junction Boxes; Conduit Bodies; Fittings; and Handhole Enclosures

Ceiling Fan Boxes

Fiberglass Round Box
Fan Rated

Thermoplastic Round Box
Fan Rated

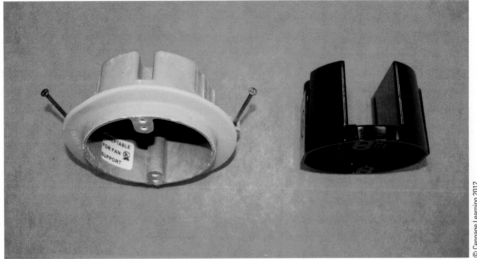

© Cengage Learning 2012

1. Ceiling fan boxes come in many styles.
2. Ceiling fan boxes must be UL listed as suitable for fan support.
3. Ceiling fan boxes in some styles support the fan independent of the box itself. In these cases, screws are used to fasten the fan directly into the stud or framing member.

NEC® Article 314, Outlet, Device, Pull, and Junction Boxes; Conduit Bodies; Fittings; and Handhole Enclosures

4" Square Boxes and Mudrings

4" Square
Thermoplastic

Single-Gang Mudring
Phenolic

2-Gang Mudring
Thermoplastic

© Cengage Learning 2012

1. Nonmetallic 4" square boxes may be used in place of other boxes for various reasons:
   a. More volume is needed.
   b. Other boxes are too deep.
   c. It is desirable to mount device horizontally.

2. Nonmetallic 4" square boxes are used with mudrings (plaster rings) to bring the edge of the box out to the finished wall. These plaster rings come in various styles, thicknesses, and materials:
   a. 1 gang
   b. 2 gang
   c. 3" round

NEC® Article 314, Outlet, Device, Pull, and Junction Boxes; Conduit Bodies; Fittings; and Handhole Enclosures

Cut-in Boxes

Single-Gang Cut-in
Fiberglass

Single-Gang Cut-in
Thermoplastic

Round Cut-in
Phenolic

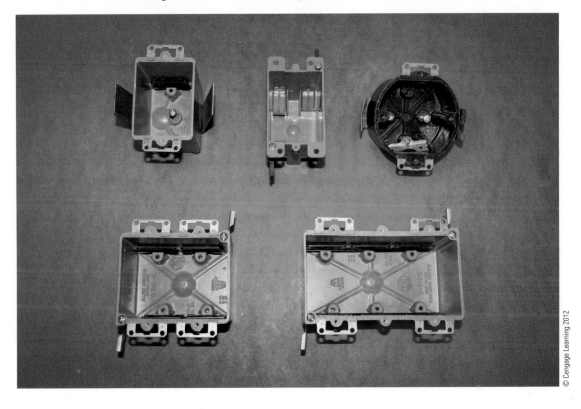

© Cengage Learning 2012

2-Gang Cut-in
Fiberglass

3-Gang Cut-in
Fiberglass

1.   Nonmetallic cut-in boxes (old work boxes) are used to add outlets to a finished wall.

NEC® Article 314, Outlet, Device, Pull, and Junction Boxes; Conduit Bodies; Fittings; and Handhole Enclosures

Box Extenders

Round Box Extender   **Single-Gang Box Extender**

© Cengage Learning 2012

1. Box extenders are used to extend a box when the front edge of the box is not flush with the finished surface.

2. Box extenders are available in round, single gang, and 2 gang. The single-gang extender can be used to extend 2-gang boxes, 3-gang boxes, and so forth.

NEC® Article 314, Outlet, Device, Pull, and Junction Boxes; Conduit Bodies; Fittings; and Handhole Enclosures

FS (Weatherproof) Boxes

Single-Gang FS Box
Internal Hubs

2-Gang FS Box
External Hubs

© Cengage Learning 2012

1. FS boxes are weatherproof device boxes that, with the appropriate cover and seal, can be mounted in wet locations.

2. FD boxes are deep weatherproof device boxes.

3. FS boxes are available with hubs for the attachment of rigid PVC.

NEC® Article 314, Outlet, Device, Pull, and Junction Boxes; Conduit Bodies; Fittings; and Handhole Enclosures

Junction (Pull) Box

Nonmetallic Junction Box (Pull Box)

1. Nonmetallic junction boxes are available in various sizes and depths.

2. Nonmetallic junction boxes can be mounted outdoors and in wet locations, with the appropriate ratings.

3. Nonmetallic junction boxes are used to splice conductors and mount components, and as a pull point.

NEC® Article 314, Outlet, Device, Pull, and Junction Boxes; Conduit Bodies; Fittings; and Handhole Enclosures

Floor Outlet Box

Round Nonmetallic Floor Outlet Box and Outlet Assembly

© Cengage Learning 2012

1. Floor outlet boxes provide an enclosure for receptacles in concrete floors.

2. Floor outlet boxes and covers must be appropriately listed for use in floors.

3. Floor outlet boxes are available in various depths, sizes, and shapes (round, rectangular).

4. Floor outlet boxes are available with dividers to separate line voltage conductors from communication cables.

NEC® Article 314, Outlet, Device, Pull, and Junction Boxes; Conduit Bodies; Fittings; and Handhole Enclosures

Communications/Low-Voltage Boxes and Rings

2-Gang Box with Divider	Single-Gang Ring Nail-on	Single-Gang Ring Cut-in	Single-Gang Ring Side Bracket

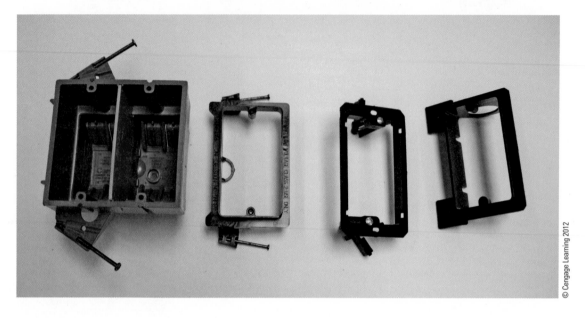

© Cengage Learning 2012

1. A 2-gang box with a divider allows line voltage on one side of the box and low voltage on the other.
2. A low-voltage ring provides an outline to cut the sheetrock around and provides holes to screw the plate into.

NEC® Article 314, Outlet, Device, Pull, and Junction Boxes; Conduit Bodies; Fittings; and Handhole Enclosures

Surface-Mounted Box

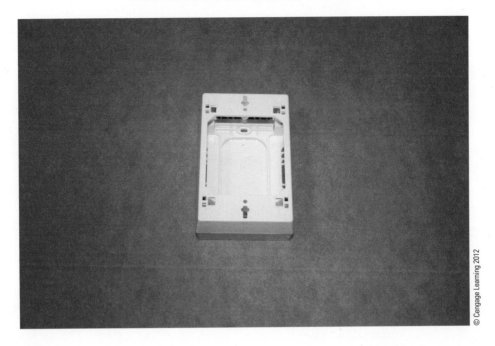

© Cengage Learning 2012

1. Surface-mounted nonmetallic boxes are used to add outlets in finished rooms.
2. They are often used in conjunction with surface nonmetallic raceways.
3. Surface-mounted nonmetallic boxes are typically used for low-voltage or communication outlets.
4. Surface-mounted nonmetallic boxes are available with an adhesive back or may be fastened with screws.
5. If screwed in place, surface-mounted nonmetallic boxes may be listed for use up to 600V.

NEC®Article 314, Outlet, Device, Pull, and Junction Boxes; Conduit Bodies; Fittings; and Handhole Enclosures

Siding Blocks

2-Piece Luminaire
Siding Block

Doorbell Button
Siding Block

Receptacle Siding Block
with In-Use Cover

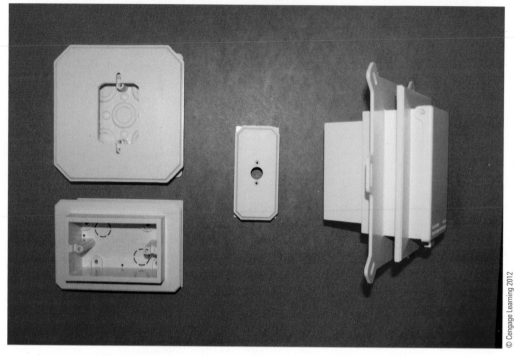

© Cengage Learning 2012

Receptacle Siding Block

1. Siding blocks come in many styles.

2. The siding blocks pictured above are listed for use as junction boxes for conductors. These types of blocks have holes that fixtures and devices screw into.

3. Siding blocks may not be rated to be used as junction boxes and require a box to be installed behind or inside them.

4. Siding blocks may have two pieces (e.g., the luminaire-type block pictured above): a back plate that is mounted to the sheathing of the house and provides an edge to side around, and a front plate that acts as a decorative cover.

5. Siding blocks are available with a built-in "in-use" cover.

NEC® Article 314, Outlet, Device, Pull, and Junction Boxes; Conduit Bodies; Fittings; and Handhole Enclosures

In-Use Weatherproof Cover

© Cengage Learning 2012

1. In-use covers are now required on recceptacles in wet locations.
2. In-use covers are designed to keep water from entering the device while a cord is plugged in.

Meter Socket Siding Block

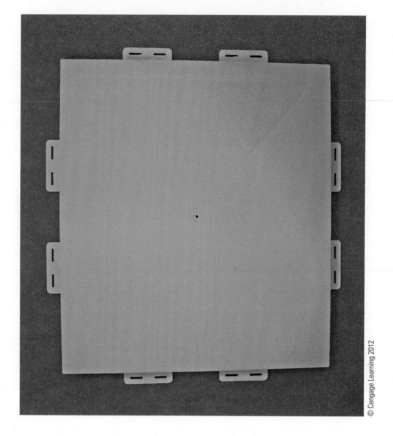

© Cengage Learning 2012

1. A meter socket siding block provides a mounting surface for a meter socket.
   a. Siding fits into channel around the block.
   b. The meter socket siding block allows siding to expand and contract without being pinched by the mounting screws of meter socket.
2. Meter socket siding blocks are available in various sizes.

CHAPTER

Metallic Boxes

INTRODUCTION

Boxes are used as a place to pull wires, splice wires, mount devices, mount lights, and so on. They can be divided into two categories: nonmetallic boxes and metallic boxes. This chapter will focus on metallic boxes.

Metallic boxes can be divided into a few general categories: device boxes, square boxes, octagon or round boxes, and pull boxes. Metallic device boxes, like the name implies, are intended to mount devices and have holes tapped for 6-32 screws. Square boxes may be used to house a device, hang a luminaire, or as a junction box. If a square metallic box is to be used with a device or luminaire, a mudring will be required to bring the box flush with the finished wall surface and provide mounting holes that align with the device or luminaire strap. Octagon and round boxes are typically used to mount a luminaire. Square, round, and octagon boxes all have holes tapped for 8-32 screws to allow the necessary support for hanging luminaires.

Metallic boxes are typically used with metal raceways; however this isn't always the case. Non-metallic cables and raceways may be used with metallic boxes. One difference between nonmetallic and metallic boxes is that most metallic boxes have round knockouts sized according to raceway trade sizes, rather than small openings, for cables to be inserted into. The knockouts may be sized specifically for one size raceway or may have concentric or eccentric knockouts that allow for several size raceways to fit into one knockout (Figures 3-1 and 3-2). Metallic boxes with small openings for cables are available if the need arises.

Larger boxes, better known as pull boxes or enclosures, are available with or without knockouts. Many times it is preferable to order a box without knockouts and make your own hole exactly where you want rather than trying to work with what is already there.

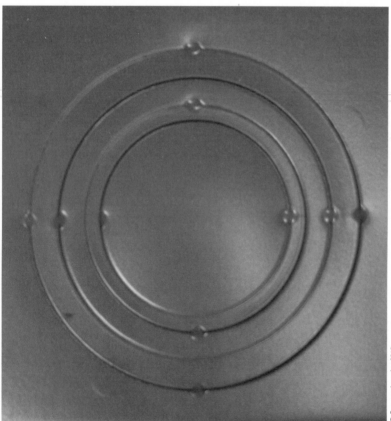

© Cengage Learning 2012

FIGURE 3-1 CONCENTRIC KNOCKOUT

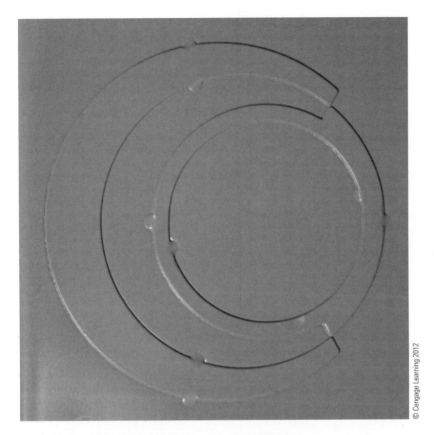

© Cengage Learning 2012

FIGURE 3-2 ECCENTRIC KNOCKOUT

Metallic boxes will have a maximum volume in cubic inches that they can house. Items that must be figured into the calculation include conductors, devices, internal clamps, fixture studs, and hickeys. *Article 314.16* of the *NEC®* has detailed instructions on how to perform box fill calculations. Smaller metallic boxes are generally built in standard sizes, which keeps the volume standard from one manufacturer to another. *NEC® Table 314.16(A)* lists the volume for smaller metallic boxes. Any box that is built with a volume that varies from those given in the *NEC®* table is required to have that volume stamped on the box. The volume of larger metallic boxes can be found by simply multiplying the dimensions of the box. Boxes with conductors #4 or larger are sized using the rules for pull box sizing found in *Article 314.28* of the *NEC®* rather *than Article 314.16*.

The National Electrical Manufacturers Association (NEMA) creates standards to which many electrical products are built to conform. By purchasing an item built to a NEMA standard, you can be sure that it was designed to perform the particular task at hand. Electrical boxes (enclosures) are built according to these standards. For example, if a box is built to meet the NEMA 3R standard, it has been designed to be installed outdoors where it will be exposed to the rain. It will protect the conductors or whatever else is mounted in the box from getting wet. *Table 110.28* of the *NEC®* lists many of the NEMA enclosure ratings and the protection they offer.

4" Square Boxes and Mudrings

4 sq × 2⅛"   4 sq × 1½"   4 sq × 1½" Side Bracket

4" sq Extension Ring   4" sq Mudring (2 gang)   4" sq Mudring (3" round)

1. Four-inch square boxes are used most often in commercial and industrial installations.
2. Four-inch square boxes are used to contain devices, splice wires, and provide pull points.
3. Four-inch square boxes are commonly referred to as *4 square* and *4 square deep*: *4 square* refers to the 1½"-deep box; *4 square deep* refers to the 2⅛"-deep box.
4. Four-inch square boxes are available with a side bracket that allows the boxes to be easily fastened to the framing member. There are several styles of side brackets.
5. Four-inch square boxes come with ½" and ¾" trade-size knockouts on the side and the back.
6. Four-inch square boxes require a mudring (plaster ring) when used to contain devices or to hang a fixture. Mudrings come in many depths, to allow for the thickness of the wall finish, and in three styles: single gang, 2 gang, and 3" round.
7. Four-inch square boxes that are used with a mudring in a finished wall should have the front of the 4-square box installed flush with the framing member. The front edge of the mudring should be flush with the finished wall surface.
8. Four-inch square boxes can be used with 4 square extension rings to increase the volume of a box.

NEC® Article 314, Outlet, Device, Pull, and Junction Boxes; Conduit Bodies; Fittings; and Handhole Enclosures

4¹¹/₁₆" Boxes and Mudrings

4¹¹/₁₆" × 2⅛"

4¹¹/₁₆" Mudring
(single gang)

4¹¹/₁₆" Mudring
(3" round)

© Cengage Learning 2012

1. Four-and-11/16-inch boxes are used for the same applications as 4 square boxes, but the former are larger and have more volume for wires, splices, and so on.

2. Four-and-11/16-inch boxes generally have ½" and ¾" knockouts, but also are available with 1" knockouts.

3. Four-and-11/16-inch boxes are available in various depths and with side brackets.

4. Four-and-11/16-inch extension rings are available to create additional volume.

NEC® Article 314, Outlet, Device, Pull, and Junction Boxes; Conduit Bodies; Fittings; and Handhole Enclosures

Blank Plates

4 Square Blank   4″ Round Blank Plate
(with Knockout)

© Cengage Learning 2012

1. Blank plates are used to close boxes after wires are spliced, pulled through, and so on.
2. Blank plates are available for the following boxes:
   a. 4 sq
   b. 4 $1\frac{1}{16}$"
   c. round
   d. octagon
3. Blank plates are available with prepunched knockouts in the center of the plate.

NEC® Article 314, Outlet, Device, Pull, and Junction Boxes; Conduit Bodies; Fittings; and Handhole Enclosures

4" Square Raised Covers

GFCI   Single Receptacle

Duplex Receptacle   Double Switch

© Cengage Learning 2012

1. Raised covers permit mounting a device into a surface-mounted 4 square box.
2. Raised covers are typically used in commercial applications involving a surface raceway and boxes (e.g., a shop area, mechanical room).
3. Raised covers are available for 4 11/16" boxes.

NEC® Article 314, Outlet, Device, Pull, and Junction Boxes; Conduit Bodies; Fittings; and Handhole Enclosures

Octagon, Round, and Pan Boxes

4" Octagon   3" Octagon   4" Pan Box

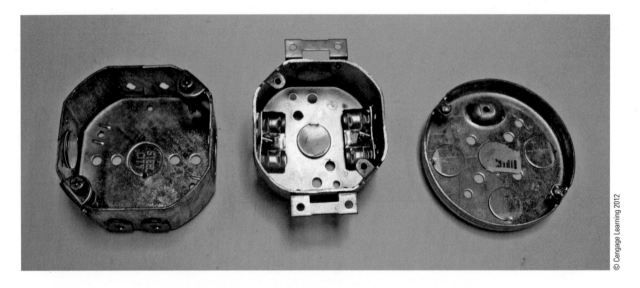

© Cengage Learning 2012

1. Octagon boxes are generally used to mount fixtures.
2. Octagon boxes are available with internal cable clamps, knockouts, or both.
3. Pan boxes are very shallow, typically only ½", ⅝", or ¾" deep. They can be used when a luminaire must be hung in a spot containing a stud or framing member. The boxes are screwed directly to the framing member.

Octagon Box with Barhanger

© Cengage Learning 2012

1. The barhanger is used when a box must be centered between two framing members.

NEC® Article 314, Outlet, Device, Pull, and Junction Boxes; Conduit Bodies; Fittings; and Handhole Enclosures

Utility Boxes (Handy Boxes) and Plates

Handy Box   Handy Plate (Blank)

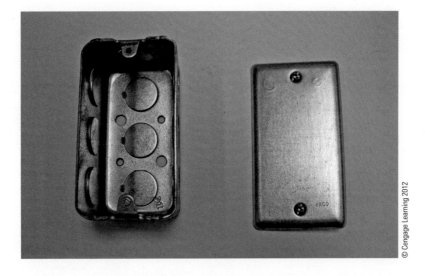

© Cengage Learning 2012

1. Handy boxes are surface-mounted device boxes that are generally used to mount a device (e.g., a switch, receptacle).

2. Handy boxes are covered with handy plates, which are available to fit almost any device.

NEC® Article 314, Outlet, Device, Pull, and Junction Boxes; Conduit Bodies; Fittings; and Handhole Enclosures

Masonry Boxes

Single-Gang Masonry Box   2-Gang Masonry Box (Gangable)

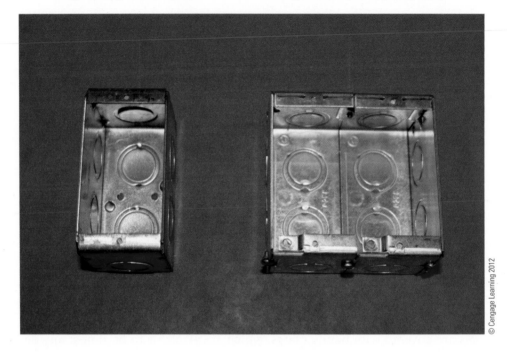

© Cengage Learning 2012

1. Masonry boxes are device boxes that are typically used in brick and masonry installations.
2. The screw holes for mounting a device are such that the attachment screws will enter the inside of the box so as not to run into the mortar or concrete that is on the outside of the box.
3. Gangable masonry boxes are available, which means that one can remove a box's side to attach one or more boxes to another, thus making a 2-gang box, 3-gang box, and so on.

NEC® 314, Outlet, Device, Pull, and Junction Boxes; Conduit Bodies; Fittings; and Handhole Enclosures

Weatherproof Boxes (Bell® Boxes) and Covers

Single-Gang WP Box	2-Gang WP Box	Round WP Extension Box

Threaded Plugs   Gasket

© Cengage Learning 2012

1. Weatherproof (WP) boxes are used to house outdoor receptacles, switches, and so on.
2. WP boxes are designed to be rain-tight.
3. WP boxes are available with ½"- or ¾"-threaded hubs.
4. WP boxes are available in several versions:
   a. single gang
   b. 2 gang
   c. 4" round
5. WP extension boxes are available to extend an existing flush mount box, or give extra volume.

NEC® 314, Outlet, Device, Pull, and Junction Boxes; Conduit Bodies; Fittings; and Handhole Enclosures

Weatherproof and In-Use Covers

| Duplex Receptacle WP Cover | Round WP Cover with ½" Hub | Metal In-Use Cover |

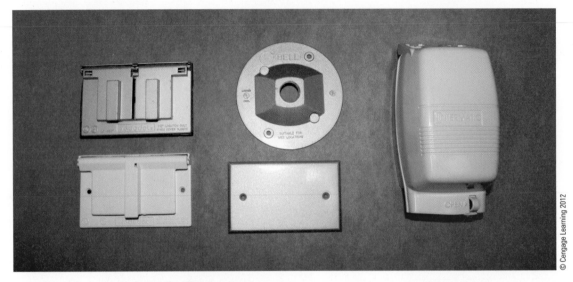

Decora (GFCI) WP Cover   Blank WP Plate

© Cengage Learning 2012

1. Weatherproof covers are available in many different styles and colors.
2. Regular flip-up weatherproof covers are permitted to be used in damp locations.
3. Receptacles in wet locations require an in-use cover so that the receptacle will stay dry even with a cord plugged in.

NEC® Article 314, Outlet, Device, Pull, and Junction Boxes; Conduit Bodies; Fittings; and Handhole Enclosures

Gangable Device Boxes (Metal Cut-in Boxes)

Single Gang   2 Gang   Mounting Bracket

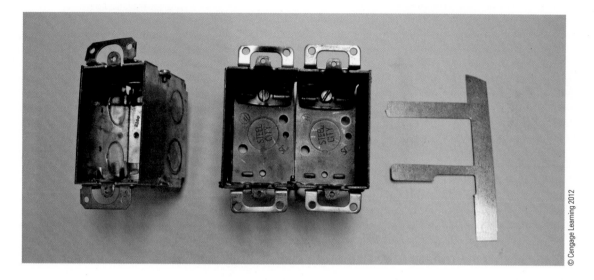

© Cengage Learning 2012

1. Metal gangable boxes are called *gangable* because one can remove a box's side to attach one or more boxes to another, thus making a 2 gang box, a 3 gang box, and so on.

2. Metal gangable boxes are available with internal cable clamps, knockouts, or both.

3. Metal gangable boxes also are available with a side bracket that allows them to be mounted to a framing member before the finished wall surface is installed.

4. Metal gangable cut-in boxes are typically used in commercial applications where an outlet must be added in a finished wall.

5. Metal gangable cut-in boxes have mounting brackets that hold the box in the wall. These brackets are sometimes called *hold its*, *F-straps*, or *submarines*.

NEC® Article 314, Outlet, Device, Pull, and Junction Boxes; Conduit Bodies; Fittings; and Handhole Enclosures

Wiremold® Boxes

Single-Gang Box   2-Gang Extension Box   Round Box

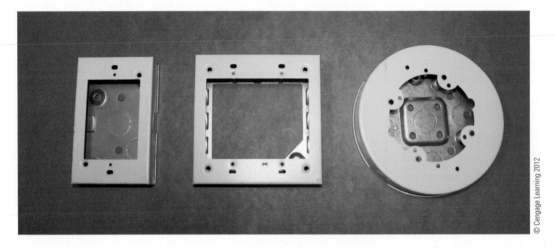

© Cengage Learning 2012

1. Wiremold® boxes are used as a decorative way to add outlets or devices on wall surfaces in finished areas.
2. Wiremold boxes are available in a variety of styles and sizes, including:
   a. single gang
   b. 2 gang
   c. 3 gang
   d. 4 round
   e. extension type (may be mounted over an existing opening, thus still allowing access to the old box)
3. Wiremold boxes can be connected directly to surface metal raceways without using a connector.

NEC® Article 314, Outlet, Device, Pull, and Junction Boxes; Conduit Bodies; Fittings; and Handhole Enclosures

Pull/Junction Box (Hoffman Box)

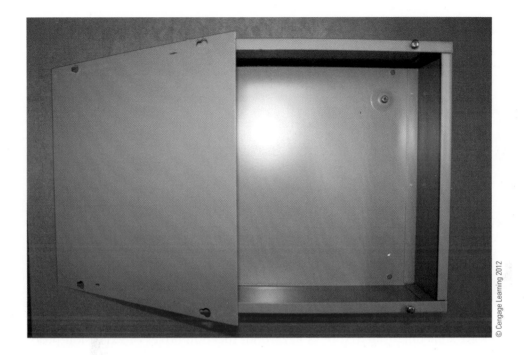

© Cengage Learning 2012

1. Pull boxes are large enclosures that are used to splice large conductors, as a place to pull wires, and to house electrical controls.

2. Pull boxes are available in many sizes and styles:
   a. various NEMA ratings
   b. with or without knockouts
   c. galvanized or painted
   d. removable cover or hinged door

NEC® Article 314, Outlet, Device, Pull, and Junction Boxes; Conduit Bodies; Fittings; and Handhole Enclosures

Cast (Explosion-Proof) Boxes

Round Box with Cover   2-Gang Box   Single-Gang Box with Switch Cover

© Cengage Learning 2012

1. Cast boxes are sometimes referred to as hazardous location boxes or explosion-proof boxes.
2. Cast boxes may be used in hazardous locations where ignitable gases, dusts, or fibers are present.
3. Explosion-proof boxes will prevent an arc inside the box from escaping and igniting combustible gases, dusts, or fibers.
4. Explosion-proof boxes have threaded hubs and are typically used with intermediate metal conduit or rigid metal conduit.

NEC® Article 314, Outlet, Device, Pull, and Junction Boxes; Conduit Bodies; Fittings; and Handhole Enclosures

CHAPTER

RACEWAYS

INTRODUCTION

Raceways provide protection for conductors and allow them to be easily installed and removed. The *National Electrical Code* definition of a raceway from *Article 100* is as follows:

> *An enclosed channel of metal or nonmetallic materials designed expressly for holding wires, cables, or busbars, with additional functions as permitted in this Code. Raceways include, but are not limited to, rigid metal conduit, rigid nonmetallic conduit, intermediate metal conduit, liquidtight flexible conduit, flexible metallic tubing, flexible metal conduit, electrical nonmetallic tubing, electrical metallic tubing, underfloor raceways, cellular concrete floor raceways, cellular metal floor raceways, surface raceways, wireways, and busways.*

The majority of the raceways, like conduit and tubing, are circular; however, this is not always the case. This chapter looks at the raceways commonly used in the electrical industry.

Conductor Fill

Raceways have a maximum internal area that conductors and cables are permitted to occupy. Figure 4-1 lists the permitted fill for conduit and tubing. This percentage can be equated to a number of conductors that will fit into a specific raceway by using the tables in *Chapter 9* of the *National Electrical Code*. Exceeding this number can cause damage to the conductors during installation. Raceways that aren't circular will have the maximum percentage of fill allowed within their respective *NEC®* article.

Allowable percentage of conductor fill for conduit and tubing	
Number of conductors	**Percentage**
1	53%
2	31%
Over 2	40%
Nipples 24" or less	60%

© Cengage Learning 2012

FIGURE 4-1

Location

Care must be taken to install raceways only where they are permitted. The following are a few examples of locations where only specific types of raceways are permitted: outside, underground, hazardous locations, wet locations, imbedded in concrete, corrosive environments, where subjected to physical damage. Consideration must also be give to expansion and contraction, particularly with rigid PVC. Expansion couplings may be required to account for the expansion and contraction when a raceway is subjected to a wide range of temperature change.

Bending

Most circular raceways that aren't flexible can be bent in the field. Hand benders are typically used to bend smaller raceways, whereas power benders are used for larger raceways. Rigid PVC, on the other hand, must be heated to bend. Once PVC is heated, it becomes flexible and can be formed into the desired shape. After the bend is made, it is allowed to cool and will retain its new shape.

It is important to keep in mind that there is a maximum of 360 degrees of bend in a run before a junction or pull box will be required. The greater number of bends, the harder it is to pull wire into the raceway. If there is more than 360 degrees in bend, the amount of force necessary to pull the conductors through will likely damage the conductors.

Electrical Metallic Tubing (Thinwall)

© Cengage Learning 2012

Definition: An unthreaded thinwall raceway of circular cross section designed for the physical protection and routing of conductors and cables and for use as an equipment grounding conductor when installed utilizing appropriate fittings. EMT is generally made of steel (ferrous) with protective coatings or aluminum (nonferrous). *NEC® 358.2*

1. Electrical metal tubing (EMT) is the most common raceway system used on commercial jobs.
2. EMT is available in many trade sizes: ½", ¾", 1", 1¼", 1½", 2", 2½", 3", 3½", and 4".
3. EMT comes in 10' lengths.
4. With the proper fittings, EMT can be used for exterior applications.
5. EMT up to size 1¼" can be bent with a hand bender.

NEC® Article 358, Electrical Metallic Tubing

Rigid Metallic Conduit (Rigid)

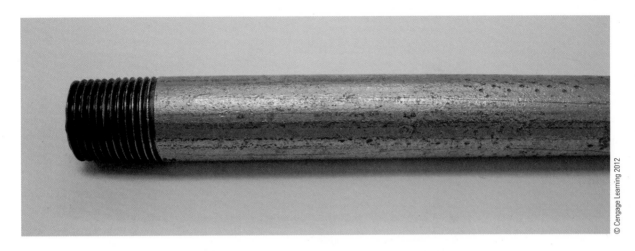

© Cengage Learning 2012

Definition: A threadable raceway of circular cross section designed for the physical protection and routing of conductors and cables and for use as an equipment grounding conductor when installed with its integral or associated coupling and appropriate fittings. RMC is generally made of steel (ferrous) with protective coatings or aluminum (nonferrous). Special use types are silicon bronze and stainless steel. *NEC® 344.2*

1. Rigid metallic conduit (RMC) is commonly used in industrial applications and some commercial applications.
2. RMC has the thickest wall of the metallic raceways. It is rated for hazardous locations, and can handle more physical abuse than other types of raceways.
3. RMC is a threaded raceway, so it must be field threaded in any place it is cut.
4. RMC must be spun together to create pipe runs, so installing RMC requires more planning than other types of raceways.
5. RMC is available in many trade sizes: ½", ¾", 1", 1¼", 1½", 2", 2½", 3", 3½", 4", 5", and 6".
6. RMC generally comes in 10' lengths with a coupling on one side and a plastic cover to protect the threads on the other side.
7. RMC comes with a colored plastic cover to indicate the trade size:
    a. black, on conduits with a "2" in the denominator: ½, 1½, 2½, 3½
    b. red, on conduits with a "4" in the denominator: ¾, 1¼
    c. blue, on conduits with whole numbers: 1, 2, 3, 4, 5, 6
8. RMC is available in steel and aluminum.

NEC® Article 344, Rigid Metal Conduit

PVC-Coated Rigid Metallic Conduit (Rob Roy)

© Cengage Learning 2012

1. PVC-coated rigid metallic conduit (RMC) has a nonmetallic coating on the outside and the inside of the raceway.
2. PVC-coated RMC is used when there may be corrosive conditions that would destroy uncoated RMC.
3. PVC-coated RMC uses couplings with a PVC coating.
4. PVC-coated RMC must have its finish integrity restored with a paste if the raceway is nicked or scratched.
5. PVC-coated RMC comes in 10' lengths, but *does not* have a coupling on one end like RMC.

NEC® Article 344, Rigid Metal Conduit

Intermediate Metallic Conduit

Definition: A steel threadable raceway of circular cross section designed for the physical protection and routing of conductors and cables and for use as an equipment grounding conductor when installed with its integral or associated coupling and appropriate fittings. *NEC® 342.2*

1. Intermediate metallic conduit (IMC) has a wall thickness in between the thicknesses of EMT and RMC.
2. IMC can handle more abuse than EMT, but not as much as RMC.
3. IMC is threaded like RMC.
4. IMC is available in many trade sizes: ½", ¾", 1", 1¼", 1½", 2", 2½", 3", 3½", and 4".
5. IMC generally comes in 10' lengths with a coupling on one side and a plastic cover to protect the threads on the other side.
6. IMC comes with a colored plastic cover to indicate the trade size:
   a. yellow, on conduits with a "2" in the denominator: ½, 1½, 2½, 3½
   b. green, on conduits with a "4" in the denominator: ¾, 1¼
   c. orange, on conduits with whole numbers: 1, 2, 3, 4

NEC® Article 342, Intermediate Metal Conduit

Rigid Polyvinyl Chloride Conduit (PVC)

© Cengage Learning 2012

Definition: A rigid nonmetallic conduit of circular cross section, with integral or associated couplings, connectors, and fittings for the installation of electrical conductors and cables. *NEC® 352.2*

1. Rigid PVC comes in two wall thickness:
    a. schedule 40
    b. schedule 80 (thicker wall)
2. Rigid PVC is available in many trade sizes: ½", ¾", 1", 1¼", 1½", 2", 2½", 3", 3½", 4", 5", and 6".
3. Rigid PVC is bent by heating the raceway until it is flexible. After RNC cools, it holds the new shape.
4. Rigid PVC uses PVC glue couple and attach fittings.
5. Rigid PVC is easy to cut with a hacksaw, PVC cutter, or string.
6. Rigid PVC is often placed underground or in sand before concrete is poured.

Note: Care must be taken when running PVC in areas where temperatures fluctuate, to allow for expansion and contraction. A 100' run, for instance, will expand and contract 6" where temperatures swing 150°F (66.5°C).

NEC® Article 352, Rigid Polyvinyl Chloride Conduit: Type PVC

Electrical Nonmetallic Tubing (Smurf Tube)

© Cengage Learning 2012

Definition: A nonmetallic pliable corrugated raceway of circular cross section with integral or associated couplings, connectors, and fittings for the installation of electric conductors. ENT is composed of a material that is resistant to moisture and chemical atmospheres and is flame retardant. *NEC® 362.2*

1. Electrical nonmetallic tubing (ENT) is a flexible nonmetallic raceway.
2. ENT is available in many colors, but typically it is blue.
3. ENT is available in ½", ¾", and 1" trade sizes.
4. ENT can be cut with a utility knife or a PVC cutter.
5. ENT can be encased in concrete when used with approved fittings.

NEC® Article 362, Electrical Nonmetallic Tubing

Liquidtight Flexible Nonmetallic Conduit (Nonmetallic Sealtite®)

© Cengage Learning 2012

Definition: A raceway of circular cross section or various types as follows:

(1) A smooth seamless inner core and cover bonded together and having one or more reinforcement layers between the core and covers, designated as Type LFNC-A

(2) A smooth seamless inner surface with integral reinforcement within the conduit wall, designated as Type LFNC-B

(3) A corrugated internal and external surface without integral reinforcement within the conduit wall, designated as LFNC-C. *NEC® 356.2*

1. Liquidtight flexible nonmetallic conduit (LFNC) is a completely nonmetallic, flexible raceway.
2. LFNC is watertight when used with the proper connectors.
3. LFNC is typically used in outdoor residential applications that require flexibility, such as an outdoor AC unit.
4. LFNC is typically available in ½", ¾", and 1" trade sizes.
5. LFNC is typically cut with a utility knife or a PVC cutter.

NEC® Article 356, Liquidtight Flexible Nonmetallic Conduit

Liquidtight Flexible Metallic Conduit (Sealtite®)

© Cengage Learning 2012

Definition: A raceway of circular cross section having an outer liquidtight, nonmetallic, sunlight-resistant jacket over an inner flexible metal core with associated couplings, connectors, and fittings for the installation of electric conductors. *NEC® 350.2*

1. Liquidtight flexible metallic conduit (LFMC) is a flexible metallic raceway with a nonmetallic outer layer.
2. LFMC is watertight when used with the proper connectors.
3. LFMC is typically used in outdoor applications that require flexibility, such as an outdoor motor or AC unit.
4. LFMC is available in many trade sizes: ½", ¾", 1", 1¼", 1½", 2", 3", and 4".
5. LFMC can be cut with a hacksaw.

NEC® Article 350, Liquidtight Flexible Metallic Conduit

Flexible Metallic Conduit (Flex) (Greenfield)

© Cengage Learning 2012

Definition: A raceway of circular cross section made of helically wound, formed, interlocked metal strip. *NEC® 348.2*

1. Flexible metallic conduit (FMC) is a flexible metallic raceway that is typically used indoors.
2. FMC is used when a flexible connection is needed, such as a motor.
3. FMC is sometimes used to feed outlets or appliances that may not require a flexible connection, such as a fluorescent light in a ceiling grid.
4. FMC is available in many trade sizes: ½", ¾", 1", 1¼", 1½", 2", 3", and 4".
5. FMC is available in steel or aluminum.
6. FMC is typically cut with a hacksaw.

   Note: Aluminum FMC is less expensive than steel FMC, but can withstand less abuse.

NEC® Article 348, Flexible Metallic Conduit

Surface Metal Raceway (Wiremold®)

Definition: A surface metal raceway is a metal raceway that is intended to be mounted to the surface of a structure, with associated couplings, connectors, boxes, and fittings for the installation of electrical conductors. *NEC® 386.2*

1. Surface raceway or Wiremold® is a surface-mounted raceway that is more decorative than a round conduit.
2. Surface raceways are available in many sizes and styles.
3. Surface raceway is used with special boxes that allow it to connect directly to a box without a connector.

NEC® Article 386, Surface Metal Raceways

Surface Nonmetallic Raceway

© Cengage Learning 2012

Definition: A nonmetallic raceway that is intended to be mounted to the surface of a structure, with associated couplings, connectors, boxes, and fittings for the installation of electrical conductors. *NEC® 388.2*

1. Surface nonmetallic raceway is a surface-mounted raceway that is more decorative than a round conduit.
2. Surface nonmetallic raceway is typically used for low-voltage or communication wiring.
3. Surface nonmetallic raceway is available with an adhesive back or may be fastened with screws.
4. Surface nonmetallic raceway is used with special boxes that allow it to connect directly to a box without a connector.
5. Surface nonmetallic raceway is often referred to as Panduit® or Nonmetallic Wiremold.®

NEC® Article 388, Surface Nonmetallic Raceways

High-Density Polyethylene Conduit

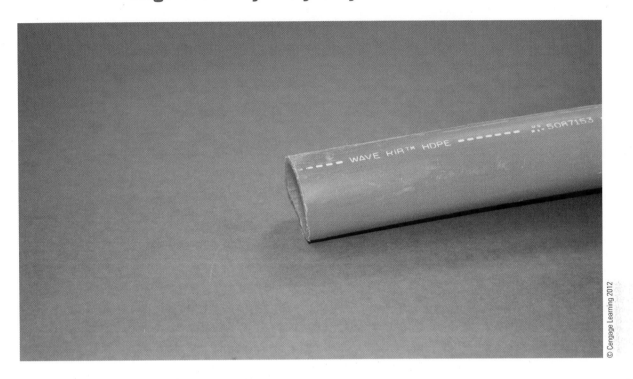

Definition: A nonmetallic raceway of circular cross section, with associated couplings, connectors, and fittings for the installation of electrical conductors. *NEC® 353.2*

1. High-density polyethylene conduit (HDPE) is designed to be installed underground.
2. HDPE is often used by communication companies as an underground raceway for fiber optic or communication cables.
3. HDPE is available in many sizes and colors.
4. HDPE is available with a factory installed pull line.
5. HDPE is available with a smooth inner wall, or ribbed to reduce the surface contact while pulling conductors or cables.

NEC® Article 353, High-Density Polyethylene Conduit

Reinforced Thermosetting Resin Conduit (RTRC)

© Cengage Learning 2012

Definition: A rigid nonmetallic conduit of circular cross section, with integral or associated couplings, connectors, and fittings for the installation of electrical conductors and cables. *NEC® 355.2*

1. RTRC is a versatile nonmetallic raceway that can be installed underground, outdoors, and in most indoor locations.
2. RTRC comes in a variety of wall thicknesses and sizes.
3. RTRC has a wide temperature rating and a very low expansion coefficient.

NEC® Article 355, Reinforced Thermosetting Resin Conduit (RTRC)

Cable Trays

Wire Mesh Cable Tray

Ladder Cable Tray

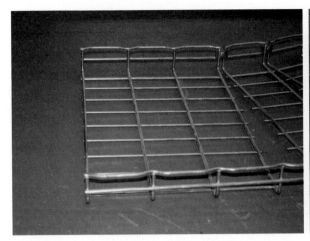

Definition: A unit or assembly of units or sections and associated fittings forming a structural system used to securely fasten or support cables and raceways. *NEC® 392.2*

1. Cable trays are used in commercial and industrial applications.
2. Cable trays make installing, removing, or adding cables very easy.
3. Cable trays come in many sizes and styles.

NEC® Article 392, Cable Trays

Wireways (Gutters)

© Cengage Learning 2012

Definition: Sheet metal troughs with hinged or removable covers for housing and protecting electric wires and cable and in which conductors are laid in place after the wireway has been installed as a complete system. *NEC® 376.2*

1. Wireways or gutters are used to make pathways that connect several boxes or cabinets.
2. Wireways may have hinged or screw-in covers.
3. Wireways are available with or without knockouts.
4. Wireways are available in many sizes and with various NEMA ratings.

NEC® Article 376, Metal Wireways

Duct Seal (Duct Mud)

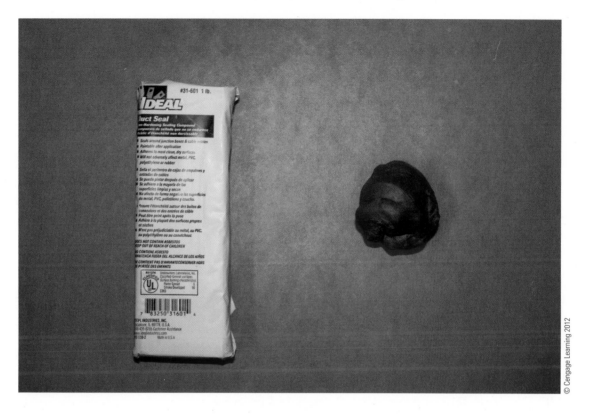

1. Duct seal is used to seal raceways that travel between areas with different temperatures, for example, those passing from the interior of a building to the exterior of a building.
2. Duct seal is malleable and can be formed and pushed onto the raceway or fitting to provide a seal.
3. Duct seal becomes very hard to work with when cold and will have to be warmed up when working in cold environments.

CHAPTER

FITTINGS

INTRODUCTION

Within the electrical industry, many fittings are used to join, fasten, and connect the various wiring methods together. *Article 100* of the *NEC®* defines a fitting as *An accessory such as a locknut, bushing, or other part of a wiring system that is intended primarily to perform a mechanical rather than an electrical function.* This chapter addresses some of the commonly used fittings in the electrical industry.

Connectors and Couplings

Connectors and couplings are fittings that are commonly used with raceways. Connectors are used to connect raceways to boxes or enclosures, whereas couplings are used to connect sections of raceway together.

Connectors and couplings must be used in the manner for which they were designed. For example, electrical metallic tubing connectors can only be used with electrical metallic tubing. Another consideration is the location of the fittings. When using connectors or couplings outdoors or in wet locations, connectors and couplings must be listed for use in wet locations; if the fitting will be in the concrete, it must be listed for use in concrete; and so on.

Fittings must be installed securely. With some fittings, this means securely tightening a set screw with a screwdriver, whereas others will require the use of a wrench to tighten them. Many times a fitting is being used to provide an equipment grounding path. If the fitting isn't fastened securely, it will jeopardize the integrity of the ground.

Connectors for flexible raceways are available straight or with an angle. Since angled connectors have such a sharp turn at the end, they will often have a cover that is removable on one side to make the installation of the conductors easier. The *National Electrical Code®* prohibits concealing angled connectors in the wall, as the sharp angle will not allow for future access to the conductors.

Nipples

Nipples are fittings or short pieces of raceway used to connect two boxes or enclosures together. The *NEC®* defines a nipple as a section of raceway 24 inches or less in length.

Conduit Bodies

Conduit bodies are used as a transition point to change the direction of a raceway and or a pull point for conductors. They will have a removable cover to allow for the transition of the conductors. Conduit bodies will typically have a cubic-inch rating, and provided it isn't exceeded will allow for conductors to be spliced. Convertible conduit bodies have removable and interchangeable hubs that allow for one conduit body to be configured to meet the need of most installations.

Electrical Metallic Tubing Fittings

Offset EMT Connector
(Die Cast) (Set Screw)

EMT Connector
(Die Cast) (Set Screw)

EMT Coupling
(Steel) (Set Screw)

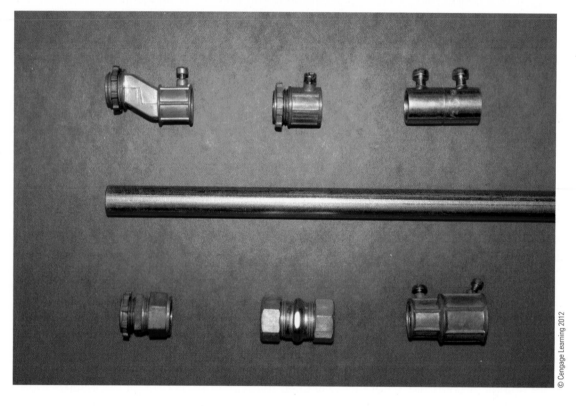

© Cengage Learning 2012

EMT Connector
(Die Cast) (Compression)

EMT Coupling
(Steel) (Compression)

EMT to FMC Coupling (Changeover)
(Die Cast) (Set Screw)

1. Electrical metallic tubing (EMT) connectors are used to attach EMT boxes or fittings.
2. Offset EMT connectors are used to attach EMT to a box without bending a box offset.
3. EMT couplings are used to couple pieces of EMT together.
4. Changeovers are used to change from EMT to flexible metallic conduit without a junction box, such as when changing to flex to feed a motor.
5. EMT fittings are available in die cast or steel, but steel fittings are more durable than die cast.
6. EMT fittings are available in set screw or compression. If properly rated, compression fittings can be used in a wet location (outdoors).

Rigid Metallic Conduit Fittings

RMC Coupling   RMC Set Screw Connector

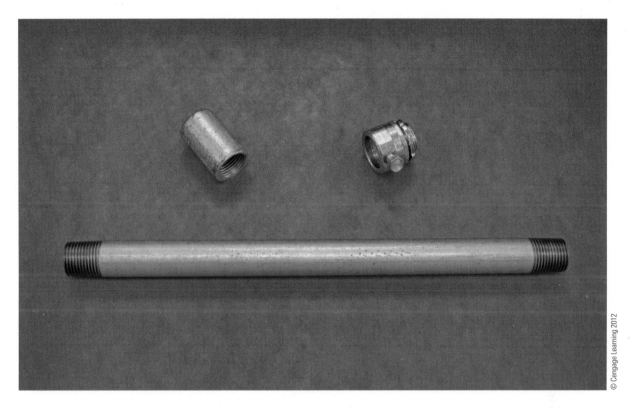

© Cengage Learning 2012

1. Rigid metallic conduit couplings join two raceways.
2. Ten-foot lengths of RMC come with a coupling on one end.
3. Set screw RMC connectors typically are used only in the event a pipe cannot be threaded (RMC is almost always threaded). A set screw connection is not watertight, nor can it be used in a hazardous location.

PVC Coated Rigid Metallic Conduit Fittings

PVC Coated Rigid Coupling PVC Coated Hub

© Cengage Learning 2012

1. PVC coated rigid couplings have a PVC boot that seals against the PVC coating of the raceway.
2. PVC coated hubs connect to a knockout or hub and provide a threaded hub that has a PVC boot that seals against the PVC coating of the raceway as well as the enclosure.

Rigid PVC Fittings

Male Adapter (TA)
(Terminal Adapter)

FA
Female Adapter

Glue-in TA
(Box Adapter)

Coupling

1. Terminal Adapters allow PVC to be glued into one side, and have male threads on the other side. They are used to connect PVC to a box or enclosure.
2. Female adapters allow PVC to be glued into one side, and have female threads in one side. They are typically used to connect a threaded raceway or connector to PVC.
3. PVC glue-in terminal adapters can be used instead of threaded terminal adapters. As their name suggests, glue-in adapters glue into a coupling or into the bell end of PVC.

Expansion Coupling

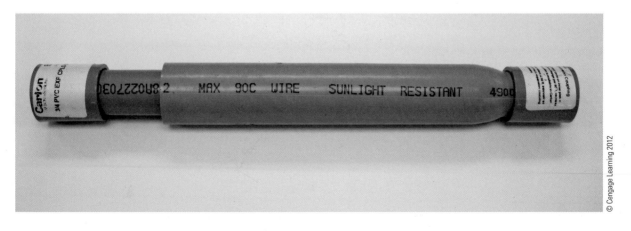

1. An expansion coupling is used where it is necessary to allow for expansion and contraction.

Electrical Nonmetallic Tubing Fittings

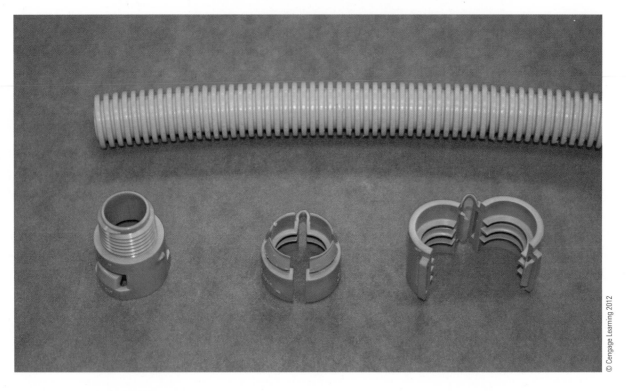

ENT Connector
(Locknut Style)

ENT Connector
(Snap-in Style)

ENT Coupling

1. Electrical nonmetallic tubing (ENT) connectors connect ENT to a box or other piece of equipment.
2. ENT couplings connect two raceways.
3. ENT connectors and couplings are available in locknut or snap-in types.

Liquidtight Flexible Nonmetallic Conduit (Sealtite®) Fittings

Straight LFNC Connector   0–90 Degree LFNC Connector

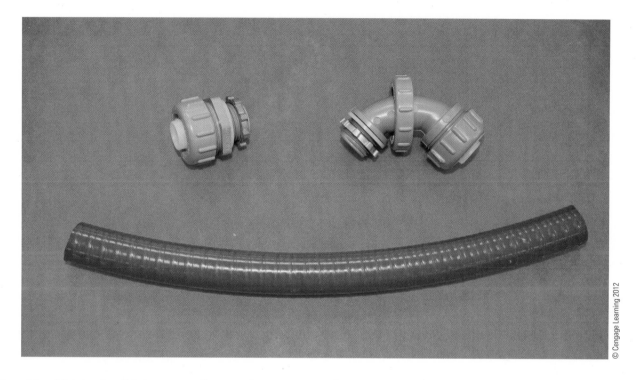

1. Liquidtight flexible nonmetallic conduit (LFNC) connectors provide a watertight seal between the raceway and the equipment or box to which it is connected.

2. LFNC connectors can be straight, have a 45-degree angle, a 90-degree angle, or be adjustable from 0 to 90 degrees.

Liquidtight Flexible Metallic Conduit (Sealtite®) Fittings

Straight LFMC Connector   90-Degree LFMC Connector

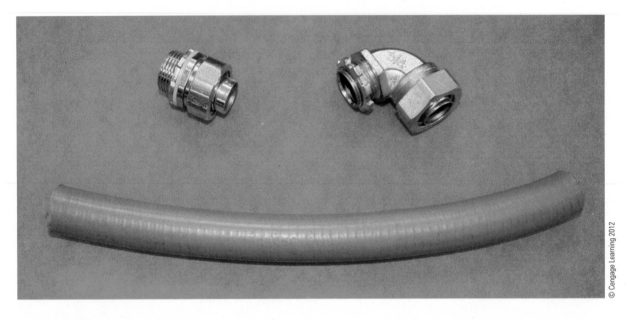

© Cengage Learning 2012

1. Liquidtight flexible metallic conduit (LFMC) fittings are metal, to bond the metallic inside of the raceway to a box or enclosure.

2. LFMC connectors provide a watertight seal between the raceway and the equipment or box to which they are connected.

3. LFNC connectors can be straight, have a 45-degree angle, or have a 90-degree angle. In some cases, a 90-degree or 45-degree connector places less stress on a raceway than does a straight connector.

Flexible Metallic Conduit Fittings

90-Degree Flex Connector (Two Screws)   Flex Connector (One Screw)

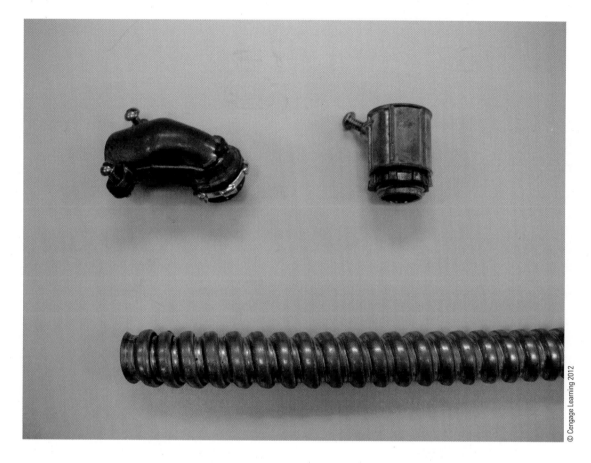

© Cengage Learning 2012

1. Flexible metallic conduit (FMC) connectors are used to connect flexible metallic tubing to boxes, fittings, or other equipment.
2. FMC connectors are available with one screw, which screws into the FMC or with two screws, which tighten a plate into the FMC.
3. FMC connectors can be straight, have a 45-degree angle, or have a 90-degree angle. In some cases, a 90-degree or 45-degree connector places less stress on a raceway than does a straight connector.

Armored Cable/Metal-Clad Cable Connectors

Straight Connector
One Screw

90-Degree Connector
Two Screws

1. Armored cable (AC) and metal-clad (MC) cable connectors are used to connect MC or AC cable to boxes, fittings, or other equipment.
2. AC and MC cable connectors require a bushing to be installed in the cable to prevent a fault between the sheath and the conductors.
3. AC and MC cable connectors are available with one screw, which screws into the cable, or with two screws, which tighten a plate into the cable.
4. AC and MC cable connectors may be straight, have a 45-degree angle, or have a 90-degree angle. In some cases, a 90-degree or 45-degree connector places less stress on the cable than does a straight connector.

Rubber Cord Connectors

Nonmetallic Cord Connector   Metallic Cord Connector

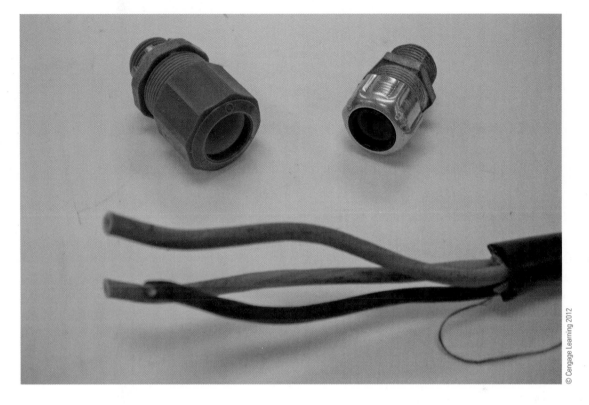

© Cengage Learning 2012

1. Rubber cord connectors are used to connect a portable cord to a box or piece of equipment.
2. Rubber cord connectors are available in many sizes and may be metallic or nonmetallic.

Strain-Relief Connectors

Strain-Relief Rubber Cord Connector (Kellum®)

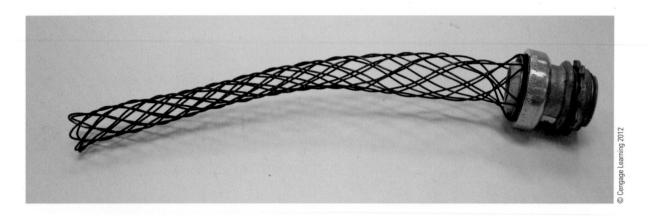

© Cengage Learning 2012

1. Strain-relief type connectors are used with hanging rubber cords. The connector's wire mesh takes the strain from the weight of the cord and transfers it from the sheath of the cable to a box or enclosure.

Surface Metal Raceway (Wiremold®) Fittings

Hooded Connector	Connector	Coupling Cover

Inside 90 (Elbow)	Side 90 (Elbow)	Backstrap

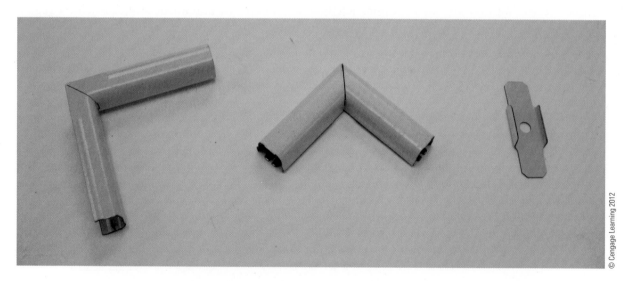

1. Wiremold® fittings are available in many styles. The pictures above represent only a small sample of the available fittings.

Miscellaneous Fittings

| Weatherproof Hub (Myers® Hub) | Threaded Nipple | Three-Piece Coupling (Erickson®) |

1. Weatherproof hubs provide a watertight entrance into an enclosure and typically are used to connect a raceway to an enclosure in an outdoor installation.
2. Threaded nipples provide a chase between boxes, enclosures, or equipment.
3. Three-piece couplings are used to splice RMC when a raceway cannot be spun on with a regular coupling. Most electricians do not use this fitting unless absolutely necessary.

| Chase Nipple | Offset Nipple | Locknut |

1. Chase nipples are used to provide a path between boxes, between a box and an enclosure, and so on.
2. Offset nipples provide a path between boxes or enclosures that do not have aligned holes.
3. Locknuts securely attach fittings to boxes or enclosures.

Conduit Seal (Seal Off)

1. Sealing fittings are used where a raceway enters or exits a hazardous location to prevent gases that may have entered the conduit from entering unclassified areas through the raceway.

Sealing Compound (Chico®)   Packing Fiber

1. The sealing compound and packing fiber is used to create a seal that will prevent gases from moving through a raceway.
2. Packing fiber is packed in the conduit seal around the conductors to prevent the sealing compound from running down into the raceway.
3. The sealing compound is mixed with water to form a paste that is poured into the conduit seal. When the compound dries, it is very similar to concrete.
4. Sealing compound is available in a premixed paste that is simply squirted into the conduit seal.

Reducing Washers	Reducing Bushing	Plastic Bushing	Grounding Bushing

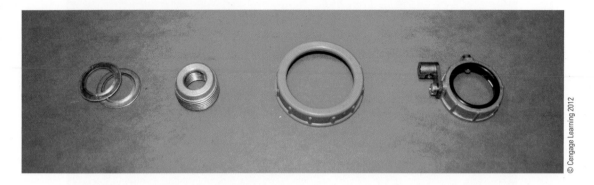

© Cengage Learning 2012

1. Reducing washers reduce the size of a knockout to allow a smaller raceway to be inserted.
2. Reducing bushings reduce the size of a threaded hub to allow a smaller raceway to be inserted.
3. Plastic bushings are installed to prevent damage to the conductors. These bushings are available in plastic, phenolic, or metal with a plastic throat.
4. Grounding bushings have a set screw to secure the bushing to the raceway and a grounding lug to allow the connector/raceway to be bonded. It also has an insulated throat to protect conductors from abrasion.

Closure Plug	Hole Closure	Weatherproof Hole Closure

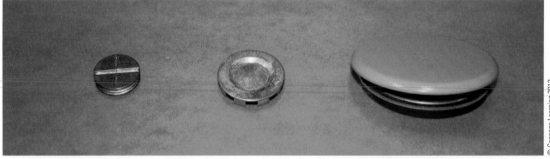

© Cengage Learning 2012

1. Closure plugs are used to plug unused holes in boxes with threaded hubs.
2. Hole closures are used to plug unused holes in boxes with knockouts.
3. Weatherproof hole closures seal open knockouts in wet locations.

Nonmetallic Sheathed Cable Fittings

Snap-in Connector (Two Cables)	Snap-in Connector (One Cable)	Two-Screw Connector	NM Cable to EMT Changeover

© Cengage Learning 2012

1. Snap-in connectors (yellow connector, above) for two cables are sometimes used on the top of panelboards.

2. Snap-in connectors (black connector, above) are typically used to connect NM to equipment such as under-cabinet lights, or surface-mounted fluorescent lights.

3. Two-screw connectors are the most commonly used type of nonmetallic sheathed cable connector. They come in two styles:

   a. locknut (shown above), which has a locknut

   b. snap-in, which does not have a locknut. Instead, snap-in connectors simply push into a box or enclosure and are locked in place by a metal tab. This type is often used in residential panels.

4. NM cable to EMT changeovers are used when nonmetallic sheathed cable is inserted into EMT (typically for physical protection).

Conduit Bodies

T   X   LR   LL

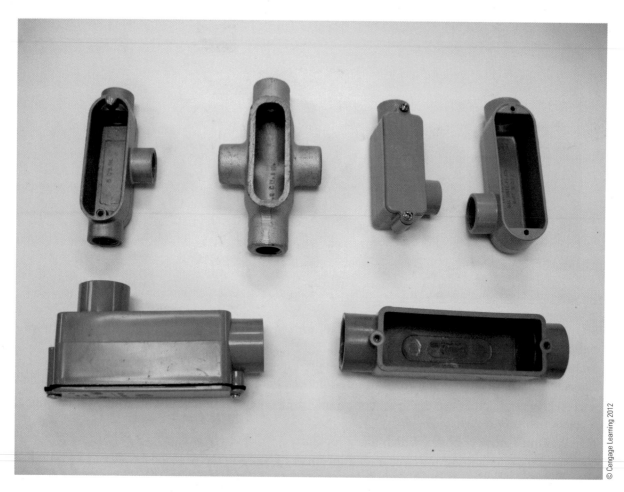

LB   C

1. Conduit bodies provide a transition point or a pull point for raceways.
2. Conduit bodies are available in a variety of materials and ratings:
   a. nonmetallic
   b. cast iron
   c. steel
3. Conduit bodies may contain spliced conductors so long as the cubic-inch capacity stamped into the fitting is not exceeded.

NEC® Article 314, Outlet, Device, Pull, and Junction Boxes; Conduit Bodies; Fittings; and Handhole Enclosures

Weatherhead (Service Head) (SE Head)

© Cengage Learning 2012

1. Weatherheads are attached to the top of a mast or raceway to provide a transition for conductors to go from inside a raceway to the outside of a raceway.
2. A weatherhead prevents rainwater from entering the raceway to which it is attached.
3. Weatherheads are available for metallic and nonmetallic raceways.

CHAPTER

FASTENING

INTRODUCTION

There are countless fasteners on the market for securing boxes, raceways, and so on. It could take an entire book to cover them all, so this chapter looks at some of the more commonly used ones. There are often several fasteners that will perform the same task, so it becomes personal preference as to the fastener that you prefer or the one required in the specifications.

Listed Use

Always install fasteners according to the manufacturers recommendations and for the purpose they were designed. The following are a few things to keep in mind:

- How much weight can it support?
- If a pilot hole is required, what diameter and depth?
- Is a washer required, and what kind?
- Does the fastener have to be corrosion resistant?
- What is the intended purpose?

Never use screws, strut, or other fasteners that will rust outdoors or in wet locations. There are many products on the market that will withstand wet and damp environments. Care must also be taken in corrosive environments.

Straps

There are many straps on the marked for fastening raceways, cables, and so forth. Be sure to use the strap for the purpose it was intended. Most conduit straps will be designed for a specific raceway, as each raceway will have a different outside diameter. For example, a ¾" EMT has a smaller diameter than a ¾" RMC. If an RMC strap were used to secure EMT, it wouldn't be tight.

Screws

Wafer Head Self-Tapping Screws   Sheetrock   Self-Drilling Screws
"Zip Screws"   Screw   "Zip Screw"   "Tek Screw"

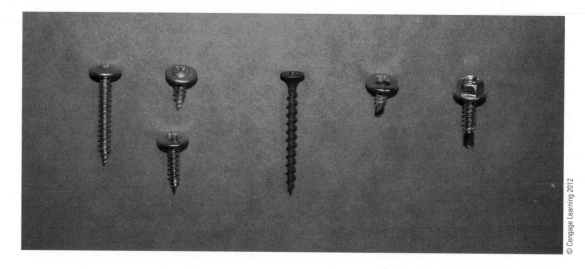

© Cengage Learning 2012

1. Wafer head screws (zip screws) are galvanized screws that are available in several lengths and with a self-tapping or self-drilling point.

2. Wafer head screws have a large head, which may eliminate the need for a washer.

3. Sheetrock screws are typically used to attach drywall to framing members; however, they are often used as a general construction screw. They do not have a coating and will rust if exposed to water or moist environments.

4. Tek screws are hex head screws that are typically used to fasten items to sheet metal.

Bolts, Nuts, and Washers

| Machine Screw | Bolt | Nut | Split Washer | Washer "Fender Washer" |

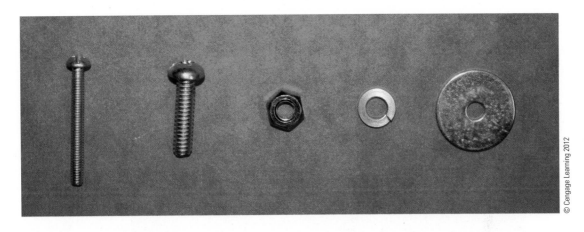

1. Machine screws are typically used to fasten devices and luminaires to electrical boxes. Commonly used sizes are as follows:

   a. 6–32   Device Boxes

   b. 8–32   Ceiling outlet boxes and 4 square boxes

   c. 10–32   Paddle fan boxes and ground screws

2. Bolts and nuts are commonly used for general fastening on construction sites.

3. Split washers are used to help keep a bolt and nut from loosening over time.

4. Washers are used to prevent the head of a screw or bolt from slipping through an opening.

Concrete Anchors

Plastic Anchor "Red Anchor"   Drive Pin "Metal"   Drive Pin "Plastic"   Concrete Screw Tapcon®   Drop-in Anchor and Setting Tool

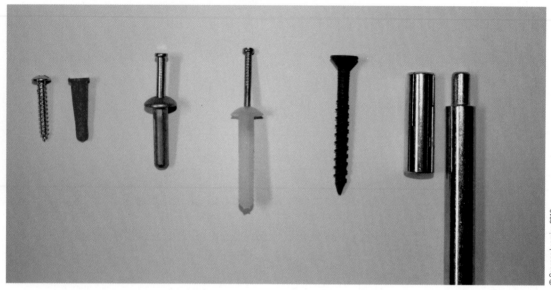

© Cengage Learning 2012

1. In order to be inserted, all of the concrete anchors in the above picture require the appropriate size hole be drilled in the concrete.

2. Red anchors are a popular anchor used when fastening raceways or boxes to concrete or masonry surfaces. The red plastic portion of the anchor is inserted into the wall, and the screw is screwed into the plastic.

3. Drive pin anchors are inserted into the hole, and then the pin is driven into the anchor with a hammer.

4. Tapcon® anchors are screwed into the hole that is drilled in the concrete.

5. Drop-in anchors must be set in place with the setting tool that expands the back side of the anchor before inserting a bolt. The hole must be the correct diameter and depth. Of the anchors shown above, this anchor offers the most support.

Drywall (Sheetrock™) Anchors

Plastic Screw-in
Anchor

Metal Screw-in
Anchor

Toggle Bolt

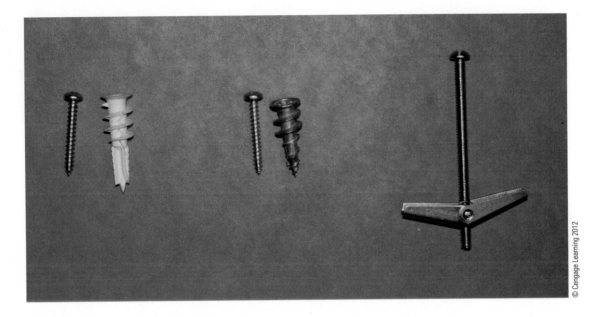

1. Screw-in anchors are used in sheetrock to anchor bathroom bar lights, raceways, boxes, and so on, and are available in plastic or metal (zinc).

2. Toggle bolts are used to fasten boxes, lights, and so forth, to sheetrock walls. A small hole is drilled in the sheetrock to allow the spring-loaded toggle to be inserted through the hole. It then expands on the backside of the sheetrock and can be secured. Toggle bolts offer the most support of the anchors shown above.

Threaded Rod

Threaded Rod   Threaded Rod Coupling

© Cengage Learning 2012

1. Threaded rod is available in various sizes and lengths.
2. Threaded rod is used to suspend raceways, pipe racks, transformers, and so on.
3. Threaded rod couplings are used to couple threaded rod pieces together.

Threaded Rod Anchors (Sammys®)

Self-Drilling with Nut   Self-Tapping   Self-Tapping

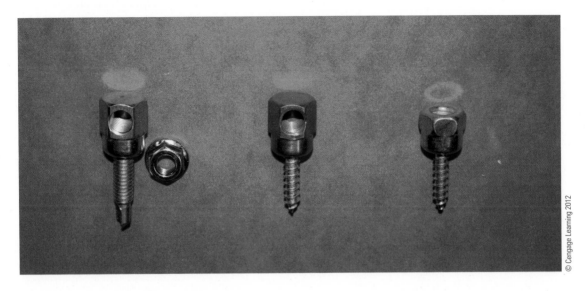

© Cengage Learning 2012

1. Threaded rod anchors can be inserted into metal or wood to provide an anchor point for threaded rod.

2. Self-drilling threaded rod anchors are available with a nut that can be inserted after the anchor has been driven through the framing member to lock it in place.

3. A driver is available that attaches to a drill and will spin free when the anchor reaches the correct depth.

Beam Clamps

Bolt-on Beam Clamp	Bolt-on Beam Clamp	Pound-on Beam Clamp with Conduit Hanger

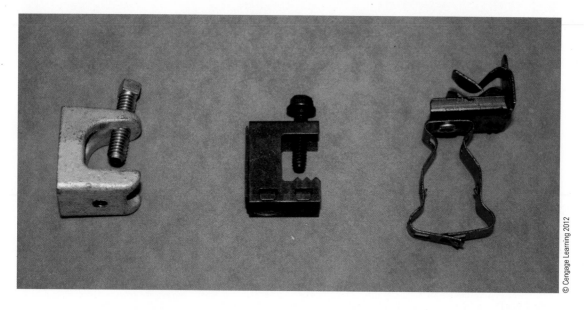

© Cengage Learning 2012

1. Beam clamps can be tightened to a beam and used to hang boxes, raceways, threaded rod, and so on.
2. Beam clamps come in many sizes and styles.
3. Beam clamps often have threaded holes on multiple sides to allow the clamp to be versatile.
4. Pound-on clamps are available with a built in conduit hanger.

Framing Channel (Unistrut®) (Strut)

Slotted Shallow Galvanized Strut	Slotted Strut	Shallow Strut

© Cengage Learning 2012

1. Strut is a metal track typically used to support electrical raceways or equipment. It is available in 10' and 20' lengths, as well as many depths and styles; the three shown above represent only a small sample of the available styles.
2. Galvanized strut resists rusting, so it is often used in wet or damp locations.
3. Strut is available with holes, slots, or a solid back.
4. Solid back strut with a cover and end plates may be used as a raceway.

NEC® Article 384, Strut-Type Channel Raceway

Framing Channel Fittings

Channel Nut   Spring Nut   Square Washer   Angle Fitting

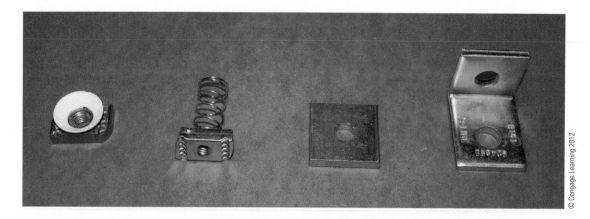

© Cengage Learning 2012

1. Channel nuts and spring nuts fit into the channel of Unistrut® and provide threaded entries for mounting boxes, straps, equipment, threaded rod, and so on.

2. Square washers are used when securing threaded rod into a spring nut/strut. A nut alone would slip through the channel, so the large square washer is used to prevent this from happening.

3. There are many different angle fittings available for building pipe racks and support frames with framing channel.

Strapping Devices

Nail-on 1-Hole Conduit Hanger
2-Hole Minerallac® Unistrut®

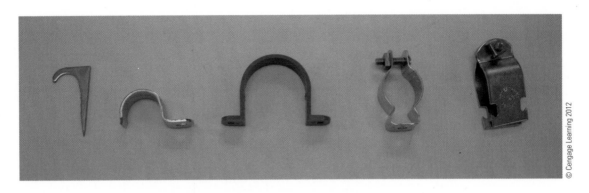

© Cengage Learning 2012

1. Nail-on straps are used to attach a raceway to a wood surface.
2. One-hole straps fasten a raceway with one screw.
3. Two-hole straps fasten a raceway using 2 screws.
4. Minerallac® straps (Minis) hold a raceway slightly away from a wall. These straps line the raceway up with the knockouts on some boxes so that the raceway can enter the box without having a box offset.
5. Unistrut® straps attach raceways to framing channels (strut).
6. Some styles of straps are available with a PVC coating to use when running PVC coated RMC.
7. One-hole straps are available with a back plate to hold a raceway away from the wall.

Note that all of the strapping devices are designed to be used with a specific type and size of raceway. Each raceway type has a different outside diameter, so, for instance, a ½" EMT strap is not the same size as a ½" RMC strap.

Cable Straps

NM Cable
Strap

Stacking
NM Cable
Strap

SER Strap

SEU Strap

Staple Gun
Cable Strap

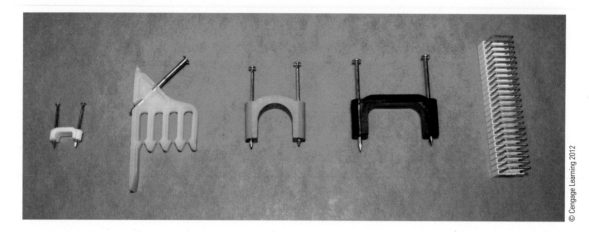

© Cengage Learning 2012

1. NM cable straps (Handy straps) are used to fasten nonmetallic sheathed cable to a wood framing member.
2. Stacking NM cable straps are used when multiple nm cables are running down a stud.
3. SER straps are used to fasten service-entrance (Type R) cable.
4. SEU straps are used to fasten service-entrance (Type U) cable.
5. Staple guns are available that will staple nm cable, communication cable, and so on, to studs. The correct staple must be inserted into the gun for the type of cable being fastened.

Nail Plates

1. Nail plates protect cables that run through studs and may be too close to the edge of the stud. This prevents sheetrock screws and trim nails from penetrating cables.

Hanging Devices

Jack Chain   Cable Hanger

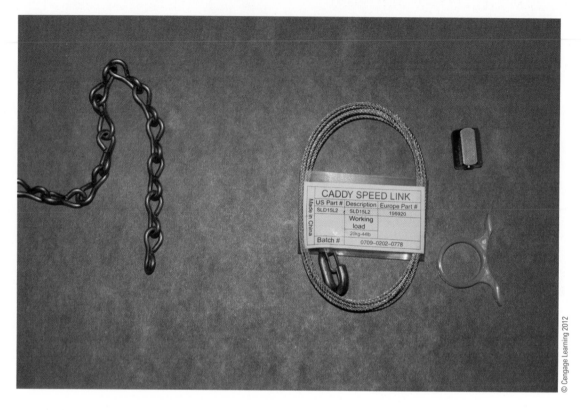

CADDY SPEED LINK

US Part #	Description	Europe Part #
SLD15L2	SLD15L2	195920
	Working load	
	20kg-44lb	
Batch #	0709-0202-0778	

Made in China

1. Jack chain and cable hangers are used to suspend luminaires and other equipment.

Cable Ties (Zip Tie) (Ty Rap®)

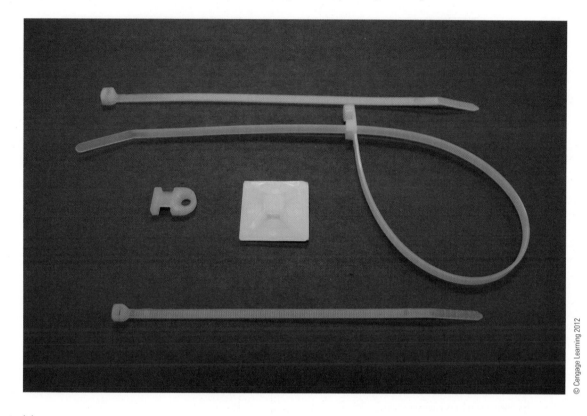

© Cengage Learning 2012

1. Cable ties come in many colors, sizes, and lengths.
2. Cable tie fasteners may have an adhesive back or fasten with a screw.
3. Cable ties have many uses such as organizing wires in enclosures or panelboards.

Roof Jack

1. Roof jacks are used to create a seal around a service mast.
2. Roof jacks are typically used on residential applications that have shingles.

SERVICE AND DISTRIBUTION EQUIPMENT

INTRODUCTION

Service and distribution equipment is a broad description of the components used in an electrical system to measure the amount of power used, distribute the energy, and provide overcurrent protection. This includes metering equipment, switchboards and panelboards, wireways, disconnects, and overcurrent protection.

Any equipment that is used as a part of an electrical service must be rated for service equipment use. It is also important to keep in mind that all electrical equipment will have maximum voltage and current ratings that must not be exceeded.

There are two types of current ratings to keep in mind when sizing service equipment and overcurrent protection, the ampere rating and the interrupting rating. The ampere rating (rated current) is the maximum amount of current that is expected to flow based on normal operation. If the rated current is exceeded, the overcurrent device will open the circuit before the excess current damages the conductor insulation. The interrupting rating (fault rating) (short-circuit current rating) is the amount of current a piece of equipment or device can safely interrupt in the event of a ground fault or phase-to-phase fault. This current will be many times greater than the rated current, potentially tens of thousands to hundreds of thousands of amps. If the equipments interrupting rating isn't at least as high as the available fault current, a fault could cause a dangerous explosion. The important thing to remember is to keep both ratings in mind when installing or replacing service equipment and overcurrent protection.

Meter Sockets

Meter Socket

Dual Meter Socket

© Cengage Learning 2012

1. A meter is used by utility companies to measure power consumption.
2. Meter sockets such as the ones above are used for line voltage or direct metering.
3. Meter sockets come in many styles and amperages, including these:
   a. standard meter socket
   b. meter socket with bypass horns, which allows utility companies to bypass a meter with jumpers if the meter needs to be removed
   c. meter socket with bypass lever, which allows utility companies to bypass a meter with a lever if the meter needs to be removed.

NEC® Article 312, Cabinets, Cutout Boxes, and Meter Socket Enclosures

Current Transformer Cabinets

CT Cabinet and Meter Socket

Current Transformers inside the Cabinet

1. Current transformer (CT) metering is used by utility companies to measure power consumption by larger services, as well as by some off-peak systems.
2. The power consumed by a building moves through the center of the current transformer, inducing a voltage and current flow that spins a meter.
3. CT cabinets are enclosures where the current transformers are housed and are the point where the service conductors coming from the utility transformer connect to the service entrance conductors of the building.
4. Off-peak CT metering systems do not always use a CT cabinet. Instead, the current transformers are housed by a large junction box, and the conductors running to the heating load are run through the CTs.

Note that in CT metering, the power consumed does not actually go through the meter socket.

NEC® Article 312, Cabinets, Cutout Boxes, and Meter Socket Enclosures

Panelboards

Three-Phase Main
Lug Panelboard

Single-Phase Main Breaker
Panelboard

 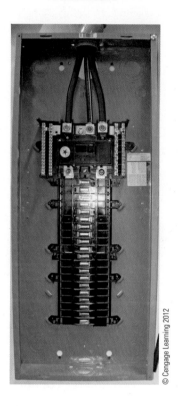

© Cengage Learning 2012

Definition: A panelboard is a single panel or group of panel units designed for assembly in the form of a single panel, including buses and automatic overcurrent devices, and equipped with or without switches for the control of light, heat, or power circuits; designed to be placed in a cabinet or cutout box placed in or against a wall, partition, or other support; and accessible only from the front. *NEC® Article 100*

1. Main-breaker panelboards contain a large OCPD to which are connected the conductors feeding the panel. This device protects the entire panel.

2. Main-lug panelboards do not have a main breaker. Instead, the conductors feeding the subpanel connect to lugs mounted on the busbars.

3. Main-lug panels are typically used as subpanels, which are protected by the OCPD in the panelboard or switchboard that feeds the subpanel.

4. Panelboards come in many sizes and amperage ratings.

NEC® Article 408, Switchboards and Panelboards

Switchboard (Switchgear)

© Cengage Learning 2012

Definition: A large single-panel, frame, or assembly of panels on which are mounted on the face, back, or both, switches, overcurrent and other protective devices, buses, and usually instruments. Switchboards are generally accessible from the rear as well as from the front and are not intended to be installed in cabinets. *NEC® Article 100*

1. Switchboards are used in large commercial and industrial buildings.

NEC® Article 408, Switchboards and Panelboards

Generator Panels

 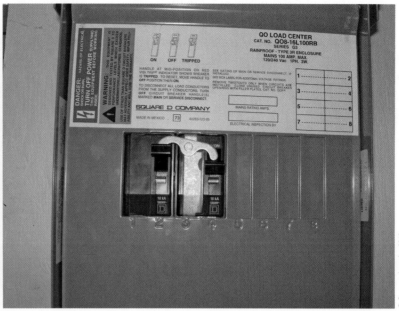

1. A generator panel is used to connect a generator to the electrical system in the case of a power outage.

2. There are many styles of generator panels; the style differentiations are based on type of feed and method of power changeover.

   a. Types of feeds include

      i. a receptacle to connect the generator cord on the front of the panel

      ii. a receptacle to connect the generator cord in a remote location (typically outside)

      iii. a direct connection.

   b. Ways to changeover power include

      i. individual switches for each load (pictured above)

      ii. one switch to changeover from utility power to generator power

      iii. an automatic transfer switch (changes over automatically).

Transformers

Transformer (Consumer Owned)
Dry Type

Transformer (Utility Owned-Pad Mounted)
Oil Filled

© Cengage Learning 2012

Transformer (Utility Owned-Pole Mounted)
Oil Filled

1. Transformers are used to increase voltage, decrease voltage, or provide electrical isolation.

2. Consumer-owned transformers (pictured above) are used to change the voltage on the consumer side of an electrical service. This transformer may step voltage up or down.

3. Utility-owned transformers (pictured above) are typically used to step the utility voltage down to the voltage used by the consumer.

NEC® Article 450, Transformers and Transformer Vaults

Disconnects (Safety Disconnects)

Disconnect   Three-Phase Fused Disconnect

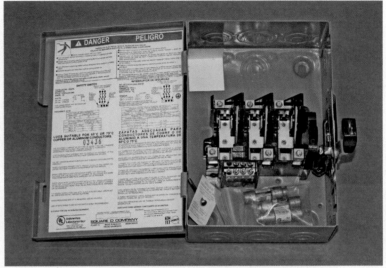

© Cengage Learning 2012

1. Disconnects are used to disconnect the power from equipment.
2. Disconnects come in many styles and ratings:
   a. fused and non fused
      i.   Fused disconnects have fuses, as in the one picture above.
      ii. Non fused disconnects do not have fuses; instead, they work like a switch.
   b. ampacities from 30 amps to 1000+ amps
   c. enclosure ratings (i.e., the NEMA rating on the enclosure) indicate which environment the disconnect is rated for.
3. Disconnects can be used as service equipment if they are designed for this purpose and marked properly.

Busways (Busducts)

© Cengage Learning 2012

Definition: A grounded metal enclosure containing factory-mounted, bare or insulated conductors, which are usually copper or aluminum bars, rods, or tubes. *NEC® Article 368.2*

1. Busways are used in large commercial and industrial locations.

2. Busways typically are run around a building to feed many loads, panels, transformers, and so on.

3. Busways are protected in their originating panel by a large fuse or circuit breaker.

4. Busways have fused disconnects, sized to protect the load to be served, attached to them at the appropriate location.

5. Busways typically contain doors placed every few feet to enable the installation of a disconnect. In the busway pictured above, note the two unused doors.

6. Busways have several disconnects. (The one pictured above has three: one on the front and two on the back side.)

NEC® Article 368, Busways

Breakers

| 1 Pole | 2 Pole | 3 Pole | Tandem 1 Pole | Bolt-in 1 Pole | AFCI 1 Pole | GFCI 1 Pole |

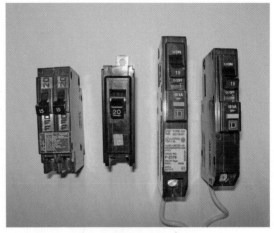

1.   One-pole breakers provide branch-circuit overcurrent and short-circuit protection for loads requiring one phase conductor: for example, a 120V receptacle outlet circuit.

2.   Two-pole breakers provide branch-circuit overcurrent and short-circuit protection for loads requiring two phase conductors: for example, a 240V baseboard heat circuit.

3.   Three-pole breakers provide branch-circuit overcurrent and short-circuit protection for loads requiring three-phase conductors: for example, a 208V 3-phase motor circuit.

4.   Tandem breakers are used when a panelboard is full and a circuit needs to be added. They take one space and turn it into two spaces. To use a tandem breaker

   a.   the panel must be rated for tandem breakers.

   b.   the maximum number of overcurrent protective devices the panel is rated for is not exceeded.

   c.   it cannot be used with two circuits of a multiwire branch circuit.

5.   Bolt-in breakers are bolted onto the busbar to create a solid connection.

6.   Arc-fault circuit-interrupter (AFCI) breakers provide protection against arc faults.

   a.   Both grounded and ungrounded conductors connect to these breakers.

   b.   Will disconnect the circuit in the event of a parallel or series arc fault.

   c.   Arc-fault protection is required on all 120-volt dwelling unit outlets in the following rooms: family rooms, dining rooms, living rooms, parlors, libraries, dens, bedrooms, sunrooms, recreation rooms, closets, hallways, or similar rooms or areas. *NEC 210.12*

7.   Ground-fault circuit-interrupter (GFCI) breakers provide protection against ground faults.

   a.   Both grounded and ungrounded conductors connect to these breakers.

   b.   Class A GFCI will disconnect the power in the event of a ground fault with a current level of 4-6 milliamps, well below a lethal current level.

   c.   *NEC Article 210.8* covers the areas where GFCI protection is required.

NEC® Article 240, Overcurrent Protection

Fuses

Cartridge Fuses
Ferrule   Knife Blade   Ferrule   Plug Fuses

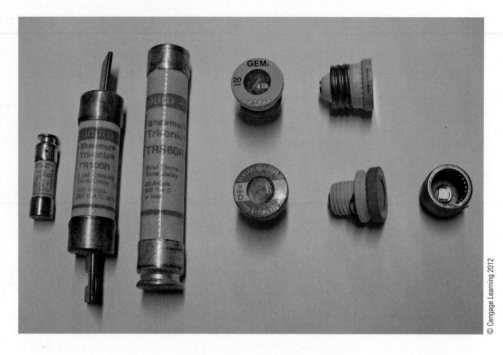

Type S (Fusestat)   Type S (Fusestat)
Plug Fuses   Adapter

1. Fuses provide overcurrent and short-circuit protection.
2. Fuses are available in ampacities from a fraction of an amp to 6000 amps.
3. Replacement fuses must have the correct voltage, amp rating, interrupting rating, and type (class-trip characteristic) as the fuses being replaced.
4. Ferrule-type cartridge fuses are used for applications of 60 amps and under.
5. Knife blade fuses are used for ampacities of more than 60 amps.
6. Plug fuses provide branch-circuit overcurrent and short-circuit protection in fuse panels. This type of protection is no longer installed in new installations; however, it is still in many older homes and buildings.
7. Type S (Fusestat) fuses are used with type S (Fusestat) adapters.
8. Fusestat adapters are installed in a plug fuse opening to restrict the fuse size. This will prevent a 20A or 30A fuse from being installed on a circuit with conductors rated for 15A.

NEC® Article 240, Overcurrent Protection

Fuse Blocks (Fuse Holders)

Fuse Holder

Fuse Holder
(Finger-safe Protection)
(Indicator Light)

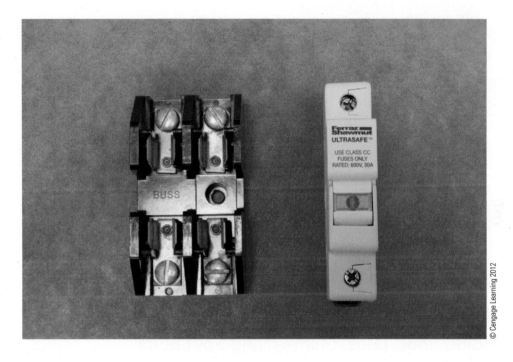

© Cengage Learning 2012

1. Fuseblocks provide a means of holding fuses and of terminating the conductors the fuse is protecting.
2. Fuseblocks are available for one, two, or three fuses.
3. Finger-safe fuseblocks allow fuses to be changed without being exposed to energized parts.
4. Indicators lights indicate whether a fuse has blown.

NEC® Article 240, Overcurrent Protection

CHAPTER

8

GROUNDING, SPLICING, AND TERMINATING

INTRODUCTION

This chapter covers grounding, splicing, and terminating devices. Grounding, splicing, and terminating are performed over and over on each jobsite. There are many devices on the market to perform these tasks. The focus of this chapter is on some of the more commonly used grounding, splicing, and terminating devices.

Aluminum versus Copper

Aluminum and copper conductors are frequently used on most jobsites. Some splicing and terminating devices are rated to be used with aluminum or copper, whereas others are not. Devices intended to be used with aluminum will be stamped with "Al." Devices intended to be used with copper will be stamped with "Cu." Devices intended to be used with aluminum or copper will be stamped with "Al/Cu." If a device is terminated with a conductor it isn't rated for, a loose connection or galvanic reaction will result over time.

Aluminum and copper conductors cannot be connected directly together as they expand and contract differently, and a galvanic reaction will occur. Some devices such as a split bolt or insulated tap allow copper and aluminum to be spliced by separating the conductors. Each conductor is terminated separately, and the device makes the electrical connection.

Ratings

All terminations and splices have maximum voltage and current ratings. It is common to have the minimum and maximum size conductor permitted stamped into the device. Exceeding these ratings can cause the connection to fail.

There is also a maximum number of conductors that are permitted to be spliced or terminated on each device. The general rule dictates that only one wire is permitted unless the device has been listed for more.

All termination and splicing devices are listed for use in specific areas. Some are designed to be in dry locations, others in wet, some underground, and so on. Be sure to follow all manufacturers' requirements, and only use the devices in locations they have been listed for.

Grounding Rods, Lugs, Clamps, and Split Bolts

© Cengage Learning 2012

1. Grounding rods are used to establish an electrical connection to earth.
2. Grounding rods are most commonly used in lengths of 8' and 10'.
3. Grounding rods are most commonly copper; however, copper ground rods typically are not solid copper, but are steel with a copper coating.
4. Grounding rods are available in ½", ⅝", and ¾" diameters.

NEC® Article 250, Grounding and Bonding

Grounding Rods, Lugs, Clamps, and Split Bolts (continued)

| Lug (2 Hole) | Ground Rod Clamp | Grounding Clamp (J-Clamp) | Grounding Clamp (J-Clamp) | Split Bolt |

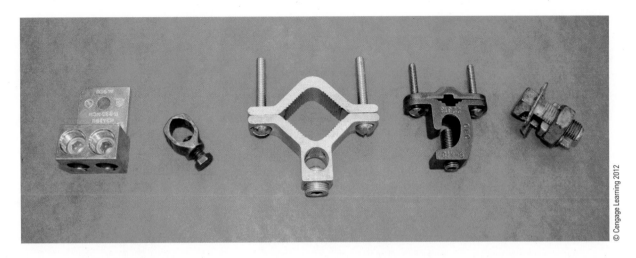

© Cengage Learning 2012

1. Lugs have multiple uses:
    a. bonding grounding conductors to enclosures or equipment
    b. terminating current-carrying conductors in meter sockets or cabinets
2. Ground-rod clamps connect the grounding electrode conductor to the grounding electrode.
3. Grounding clamps that are attached to rebar and buried in concrete must be rated appropriately.
4. J-clamps are used to connect grounding conductors to a metal rod or pipe, such as a copper water line.
5. Split bolts enable large conductors to be spliced together. Current-carrying conductors spliced with a split bolt must have the connection insulated by tape or some other approved means to prevent shorts and ground faults.
6. Split bolts, lugs, and grounding clamps are rated for the type of conductor they can terminate:
    a. Cu = copper
    b. Al = aluminum
    c. Cu/Al = dual-rated for copper and/or aluminum. This type of splice can connect aluminum to copper conductors because it contains a metal plate that separates the conductors.

Grounding Bars, Screws, Tails, Clips, and Wirenuts

Grounding Bar

Ground Screw

Ground Clip

Grounding Tail

Grounding Wirenut

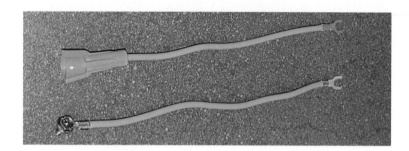

1. Grounding bars are used in panelboards or enclosures to provide a place to terminate grounding conductors.
2. Grounding screws are used to bond a conductor to a box or enclosure.
3. Grounding clips are used to bond a conductor to a box or enclosure.
4. Grounding tails are typically used to ground a device to a metallic box.
5. Grounding wirenuts have a hole in the end to allow one of the spliced conductors to pass through the wirenut to ground a device or bond the box.

NEC® Article 250, Grounding and Bonding

Splicing Wirenuts, Connectors, and Taps

Wirenuts

Push-in Wire
Connectors

Insulated
Tap

© Cengage Learning 2012

1. Wirenuts are the most common method of splicing small conductors.
2. Wirenuts are identified (on the package) according to how many conductors and what size of conductor they can splice.
3. Wirenuts are typically designed only for copper conductors. Some wirenuts, however, are available to splice copper conductors with aluminum conductors.
4. Push-in wire connectors splice wires by stripping the insulation off the conductor and pushing it into the connector.
5. Push-in wire connectors are generally limited to smaller conductors (#12 and smaller).
6. Insulated taps are available in different configurations according to the number of conductors they splice.

Underground Splicing

Butt Splice with Shrink Tube

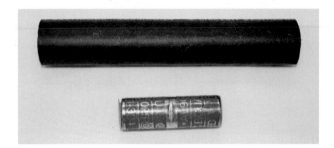

UF Splice Kit (Shrink Tube)

© Cengage Learning 2012

Cold Shrink

Wrap-Around UF Splice Kit

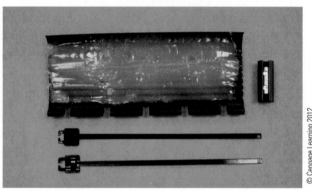

© Cengage Learning 2012

1. Butt splices are used to splice conductors (usually large conductors).
    a. A crimper must be used to mechanically connect the conductor to the splice.
    b. When heated, the shrink tube shrinks around the splice and conductor to provide a seal that can be buried underground.
2. As the name suggests, UF splice kits splice UF cable.
    a. The splice is a setscrew connection that splices together up to four conductors from each cable.
    b. When heated, the shrink tube shrinks around the splice and conductor to provide a seal that can be buried underground.
    c. Wrap-around splice kits are filled with a silicone substance that, when wrapped around the cable/splice, provides a seal that can be buried underground.
3. Cold shrink is slid over the splice, and then the plastic spiral tape is pulled out from the inside, allowing the shrink tube to collapse on the splice.

Terminals

| Crimp-on Butt Splice | Crimp-on Fork Terminal | Crimp-on Ring Terminal | Crimp-on Female Terminal | Crimp-on Male Terminal | Twist-on Fork Terminal | Twist-on Ring Terminal |

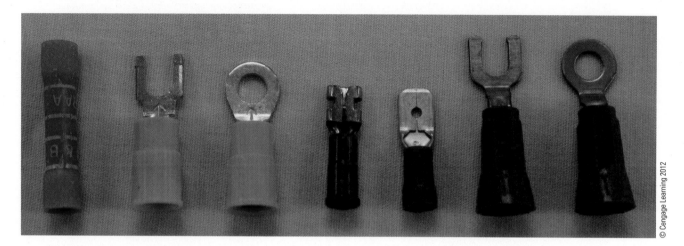

© Cengage Learning 2012

1. Ring and fork terminals are used to connect conductors to devices or equipment.
2. Crimp-on terminals and disconnects are designed for use only with stranded conductors.
3. Male and female disconnects provide a connection that can be easily changed or removed.
4. Twist-on terminals or disconnects can be used with solid, stranded, or multiple conductors.

Phase Tape

© Cengage Learning 2012

1. Phase tape is used for insulating conductors and to reidentify the color of the conductor insulation.
2. Phase tape is available in green, white, gray, black, red, blue, brown, orange, yellow, and violet.
3. When insulating a splice, splicing or rubber tape is used for the base layer, with phase tape used as the final layer.

CHAPTER

Devices

INTRODUCTION

Devices are defined by the *National Electrical Code* as *A unit of an electrical system that carries or controls electric energy as its principal function (NEC® 100).* A few common examples of devices used everyday are switches and receptacles. This chapter will cover some of the commonly used devices in the electrical industry.

This chapter will also cover a few items that aren't technically "devices" from the *NEC®* point of view, such as fire alarm and security horns and strobes, but they are included to keep similar products grouped together.

When choosing devices, the rated voltage and current must be kept in mind. Exceeding either of these values can lead to early failure or a potential hazard.

Devices come in many grades. For example, receptacle outlets are available in general-purpose grade, specification (spec) grade, and hospital grade as well as some others. Of the three listed here, general-purpose grade is the least expensive and hospital grade is the most expensive. The more expensive devices have better contact points inside, allowing them to last longer and are made of more durable materials. For example, the face on hospital-grade receptacles is made of a material that is more durable and less likely to crack than general-purpose grade.

The National Electrical Manufacturers Association (NEMA) has created standards that receptacles and plugs are built to. This ensures that all manufacturers build the receptacles to the same standard and that similarly rated plugs will fit into the receptacle.

Tamper-resistant and weather-resistant receptacles

Recent *NEC®* changes have required tamper-resistant and weather-resistant receptacles to be used in specific locations. Tamper-resistant receptacles have an internal shutter that blocks children from sticking small objects into the device. All 120-volt, 15- and 20-amp receptacles that are less than 5½ feet above the floor in dwelling units are required to be tamper resistant. In addition, 120-volt, 15- and 20-amp receptacle outlets in guest room or suites and child care areas must also be tamper resistant. Tamper-resistant receptacles will have TR stamped into the plastic, making them easy to distinguish.

Weather-resistant receptacles have been made of materials that will resist the corrosion when installed in areas exposed to damp or wet conditions. All outdoor receptacles fall into this category. The *NEC®* requires that all 120-volt 15- and 20- amp receptacles installed in wet and damp locations be of the weather resistant type. Weather-resistant receptacles will typically have WR stamped into the plastic making them easy to distinguish.

GFCI receptacles

Receptacles are available that provide ground-fault circuit interruption. Class A GFCI devices will open the circuit if a ground fault of 4–6 milliamperes or more occurs. GFCI receptacles are more often used than GFCI circuit breakers because they are less expensive. They have the capability of providing GFCI protection for all the outlets that are fed downstream from the device. GFCI receptacles will have a test and reset button on the front of the receptacle and should be tested monthly. *NEC® 210.8* lists the numerous locations that are required to have GFCI protection.

Receptacles

Duplex
Receptacle
20A

GFCI
Receptacle
15A

Single
Receptacle
30A (Dryer Outlet)

Single
Receptacle
Twist Lock

Surface-Mount
Receptacle

Clock
Receptacle

Floor
Receptacle

1. Duplex receptacles are the most commonly used receptacle.
2. Receptacles of different ampacities and voltage have different configurations to prevent their being used with the wrong plug.
3. GFCI receptacles protect people from injury from a ground fault.
4. Twist-lock receptacles require that plugs be twisted, thus locking the plugs into place.
5. Clock receptacles recess plugs allow a clock to be plugged in and hung over the receptacle without extruding from the wall.
6. Floor receptacles have a cover that can be inserted when the plug is not in use, which keeps debris out of the receptacle.

NEC® Article 406, Receptacles, Cord Connectors, and Attachment Plugs

This table is an example of a NEMA non locking plug and receptacle chart.

			15 AMPERE		20 AMPERE		30 AMPERE		50 AMPERE		60 AMPERE	
			RECEPTACLE	PLUG	RECEPTACLE	PLUG	RECEPTACLE	PLUG	RECEPTACLE	PLUG	RECEPTACLE	PLUG
2-POLE 2-WIRE	125 V	1	1-15R	1-15P			1-20P		1-30P			
	250 V	2		2-15P	2-20R	2-20P		2-30R	2-30P			
	277 V AC	3	(RESERVED FOR FUTURE CONFIGURATIONS)									
	600 V	4	(RESERVED FOR FUTURE CONFIGURATIONS)									
2-POLE 3-WIRE GROUNDING	125 V	5	5-15R	5-15P	5-20R	5-20P	5-30R	5-30P	5-50R	5-50P		
	250 V	6	6-15R	6-15P	6-20R	6-20P	6-30R	6-30P	6-50R	6-50P		
	277 V AC	7	7-15R	7-15P	7-20R	7-20P	7-30R	7-30P	7-50R	7-50P		
	347 V AC	24	24-15R	24-15P	24-20R	24-20P	24-30R	24-30P	24-50R	24-50P		
	480 V AC	8	(RESERVED FOR FUTURE CONFIGURATIONS)									
	600 V AC	9	(RESERVED FOR FUTURE CONFIGURATIONS)									
3-POLE 3-WIRE	125/ 250 V	10			10-20R	10-20P	10-30R	10-30P	10-50R	10-50P		
	3 Ø 250 V	11	11-15R	11-15P	11-20R	11-20P	11-30R	11-30P	11-50R	11-50P		
	3 Ø 480 V	12	(RESERVED FOR FUTURE CONFIGURATIONS)									
	3 Ø 600 V	13	(RESERVED FOR FUTURE CONFIGURATIONS)									
3-POLE 4-WIRE GROUNDING	125/ 250 V	14	14-15R	14-15P	14-20R	14-20P	14-30R	14-30P	14-50R	14-50P	14-60R	14-60P
	3 Ø 250 V	15	15-15R	15-15P	15-20R	15-20P	15-30R	15-30P	15-50R	15-50P	15-60R	15-60P
	3 Ø 480 V	16	(RESERVED FOR FUTURE CONFIGURATIONS)									
	3 Ø 600 V	17	(RESERVED FOR FUTURE CONFIGURATIONS)									
4-POLE 4-WIRE	3 Ø 208Y/ 120 V	18	18-15R	18-15P	18-20R	18-20P	18-30R	18-30P	18-50R	18-50P	18-60R	18-60P
	3 Ø 480Y/ 277 V	19	(RESERVED FOR FUTURE CONFIGURATIONS)									
	3 Ø 600Y/ 347 V	20	(RESERVED FOR FUTURE CONFIGURATIONS)									
4-POLE 5-WIRE GROUNDING	3 Ø 208Y/ 120 V	21	(RESERVED FOR FUTURE CONFIGURATIONS)									
	3 Ø 480Y/ 277 V	22	(RESERVED FOR FUTURE CONFIGURATIONS)									
	3 Ø 600Y/ 347 V	23	(RESERVED FOR FUTURE CONFIGURATIONS)									

This table is an example of a NEMA locking plug and receptacle chart.

			15 AMPERE		20 AMPERE		30 AMPERE		50 AMPERE		60 AMPERE	
			RECEPTACLE	PLUG	RECEPTACLE	PLUG	RECEPTACLE	PLUG	RECEPTACLE	PLUG	RECEPTACLE	PLUG
2-POLE 2-WIRE	125 V	L1	L1-15R	L1-15P								
	250 V	L2			L2-20R	L2-20P						
	277 V AC	3			(RESERVED FOR FUTURE CONFIGURATIONS)							
	600 V	4			(RESERVED FOR FUTURE CONFIGURATIONS)							
2-POLE 3-WIRE GROUNDING	125 V	5	L5-15R	L5-15P	L5-20R	L5-20P	L5-30R	L5-30P	L5-50R	L5-50P	L5-60R	L5-60P
	250 V	6	L6-15R	L6-15P	L6-20R	L6-20P	L6-30R	L6-30P	L6-50R	L6-50P	L6-60R	L6-60P
	277 V AC	7	L7-15R	L7-15P	L7-20R	L7-20P	L7-30R	L7-30P	L7-50R	L7-50P	L7-60R	L7-60P
	347 V AC	24			L24-20R	L24-20P						
	480 V AC	8			L8-20R	L8-20P	L8-30R	L8-30P	L8-50R	L8-50P	L8-60R	L8-60P
	600 V AC	9			L9-20R	L9-20P	L9-30R	L9-30P	L9-50R	L9-50P	L9-60R	L9-60P
3-POLE 3-WIRE	125/250 V	10			L10-20R	L10-20P	L10-30R	L10 30P				
	3Ø 250 V	11	L11-15R	L11-15P	L11-20R	L11-20P	L11-30R	L11-30P				
	3Ø 480 V	12			L12-20R	L12-20P	L12-30R	L12-30P				
	3Ø 600 V	13					L13-30R	L13-30P				
3-POLE 4-WIRE GROUNDING	125/250 V	14			L14-20R	L14-20P	L14-30R	L14-30P	L14-50R	L14-50P	L14-60R	L14-60P
	3Ø 250 V	15			L15-20R	L15-20P	L15-30R	L15-30P	L15-50R	L15-50P	L15-60R	L15-60P
	3Ø 480 V	16			L16-20R	L16-20P	L16-30R	L16-30P	L16-50R	L16-50P	L16-60R	L16-60P
	3Ø 600 V	17					L17-30R	L17-30P	L17-50R	L17-50P	L17-60R	L17-60P
4-POLE 4-WIRE	3Ø 208Y/120 V	18			L18-20R	L18-20P	L18-30R	L18-30P				
	3Ø 480Y/277 V	19			L19-20R	L19-20P	L19-30R	L19-30P				
	3Ø 600Y/347 V	20			L20-20R	L20-20P	L20-30R	L20-30P				
4-POLE 5-WIRE GROUNDING	3Ø 208Y/120 V	21			L21-20R	L21-20P	L21-30R	L21-30P	L21-50R	L21-50P	L21-60R	L21-60P
	3Ø 480Y/277 V	22			L22-20R	L22-20P	L22-30R	L22-30P	L22-50R	L22-50P	L22-60R	L22-60P
	3Ø 600Y/347 V	23			L23-20R	L23-20P	L23-30R	L23-30P	L23-50R	L23-50P	L23-60R	L23-60P

Receptacles (continued)

© Cengage Learning 2012

1. By looking at this receptacle, you can tell that
   a. it is 120V 15A (per the NEMA chart).
   b. it is hospital grade (if it has a green dot on the face).
   c. it is tamper resistant (the term "Tamper Resistant" or the letters "TR" will be stamped in the face). *Tamper-resistant* means that the receptacle is designed to prevent children form sticking metal objects into the slots.
   d. it is self-grounding (i.e., with a metal clip around the screw).

NEC® Article 406, Receptacles, Cord Connectors, and Attachment Plugs

1. By looking at this receptacle, you can tell that
   a. it is 120V 20A (per the NEMA chart).
   b. it is hospital grade (if it has a green dot on the face).
   c. it has an isolated ground (if it has an orange triangle on the face).
   d. it has a self-grounding yoke (i.e., a metal clip around the screw).

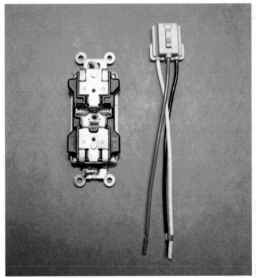

1. These receptacles can be plugged into the wiring tail provided.
2. These receptacles are designed to be replaced without interrupting power or exposing electricians to energized parts.
3. By looking at this receptacle, you can tell that
   a. it is 120V 20A (per the NEMA chart).
   b. it is hospital grade (if it has a green dot on the face).
   c. it is self-grounding (i.e., with a metal clip around the screw).

NEC® Article 406, Receptacles, Cord Connectors, and Attachment Plugs

Switches

Single-Pole
Toggle Switch

Three-Way
Toggle Switch

Decora Switch

Keyed Switch

© Cengage Learning 2012

1. Single-pole switches are used to control outlets from a single location.

2. Three-way switches are used to control outlets from two locations.

3. Four-way switches are used in conjunction with three-way switches to control outlets from more than two locations.

4. Decora switches have large rockers.

5. Lighted (pilot light) switches light up when the switch is on (closed position).

6. Illuminated switches light up when the switch is off (open position). *Note*: there must be a load attached for this switch to illuminate.

7. Double-pole switches can switch two ungrounded conductors on and off.

8. Switches are available in different grades:
   a. general purpose (residential grade)
   b. specification grade (spec grade)
   c. commercial grade
   d. industrial grade

NEC® Article 404, Switches

Miscellaneous Devices

Dimmers, Sensors, and Timers

| Rotary Dimmer | Occupancy Sensor | Rotary Timer | Digital Timer |

 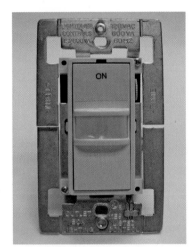  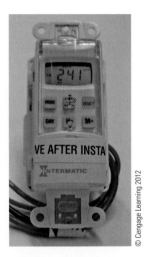

1. Dimmers are available in many styles (the rotary is shown above) and in single-pole and three-way versions.
2. Occupancy sensors sense movement and turn a lighting outlet on for a predetermined amount of time.
3. Rotary timers are spring-wound timers that shut themselves off after a set amount of time.
4. Digital timers can be set to turn on and off at predetermined times. They often have a one-week program that allows for multiple on/off settings.
5. Some digital timers require batteries.

NEC® Article 404, Switches

Miscellaneous Devices (continued)
Smoke Alarm, Ceiling Fan Controllers, Theater Lights, and Night Lights

Smoke Alarm

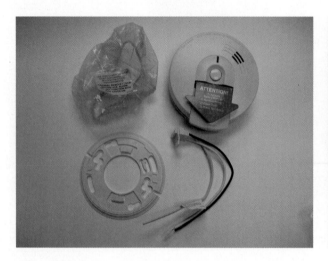

Ceiling Fan Controller

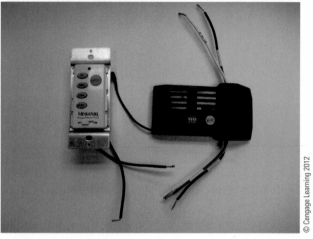

Theater Light

Night Light with Photo Sensor

© Cengage Learning 2012

1. Smoke alarms installed in new residential construction have many requirements. Different states have different codes, but several common requirements are listed below:
   a. line voltage power (120V)
   a. interconnected (i.e., if one detector senses smoke, then they all will go off)
   a. battery backup
2. Smoke alarms are ruined by construction dust, so dust covers sometimes are provided with the devices.
3. Ceiling fan controllers are designed to be used in situations where not enough wires exist between the switch and the fan to control the fan and a light separately.
4. Theater lights come in many styles and are commonly used in stairways or hallways as accent lights. The light above fits into a normal single-gang box.
5. Night lights are available with a photo sensor that turns the light on automatically when ambient light levels are low.

Fire Alarm Devices

Smoke Detector	Heat Detector	Pull Station	Audio/Visual Appliance

Door Holder · Duct Smoke Detector

1. The fire alarm devices shown above would be connected through a fire alarm panel.
2. Indicating appliances include
   a. smoke detectors, which sense the presence of smoke or other combustion particles.
   b. heat detectors, which sense unusually high temperatures. Heat detectors are typically located in kitchens or mechanical rooms where dust or smoke may be present under normal circumstances.
   c. rate-of-rise detectors, which check for a specified change in temperature in a certain time interval. These detectors are typically used in mechanical rooms or other rooms where smoke, dust, or high temperatures may be present under normal circumstances.
   d. pull stations, which enable a fire alarm system to be set off manually.
   e. duct-smoke detectors, which sample the air in ductwork to check for the presence of smoke. In the case of a fire, these detectors typically shut air handlers down and notify the fire alarm panel.
3. Notification appliances, such as audiovisual (A/V) appliances, notify occupants of the possible presence of fire.
4. Door holders hold doors open under normal circumstances. In the case of fire, they release, allowing doors to close.

NEC® Article 760, Fire Alarm Systems

Security Devices

Motion Sensor
Ceiling Mount

Motion Sensor
Wall Mount

Keypad

Magnetic Contact

Siren

1. Motion sensors can be mounted on the ceiling or the wall.
2. Motion sensors come in several types, but passive infrared (PIR) is a common one.
3. Keypads are used for programming, energizing, and de-energizing systems.
4. Magnetic contacts are installed in doors and windows to indicate they have been opened.
5. Magnetic contacts are available in normally open or normally closed contacts.
6. Security sirens are used to announce that a building's security has been breached.
7. All of the components above are connected to a security panel that monitors and maintains the system.

Faceplates

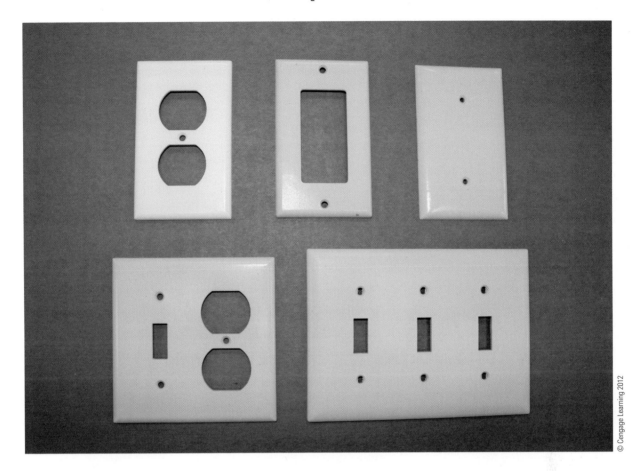

© Cengage Learning 2012

1. Faceplates are available in many colors and configurations.
2. Faceplates are available in different materials:
    a. plastic (hard)
    b. plastic (flexible, sometimes called *unbreakable*)
    c. painted steel
    d. stainless steel

Plugs and Connectors (Cord Caps)

Straight-Blade Plug   90-Degree Straight-Blade Plug

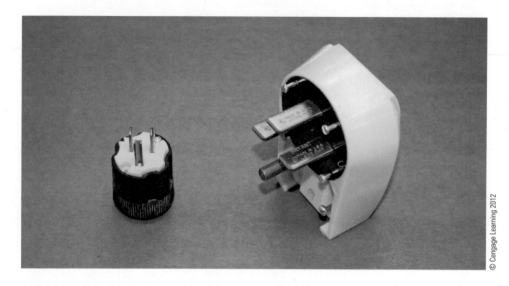

© Cengage Learning 2012

Twist-Lock Plug   Twist-Lock Connector

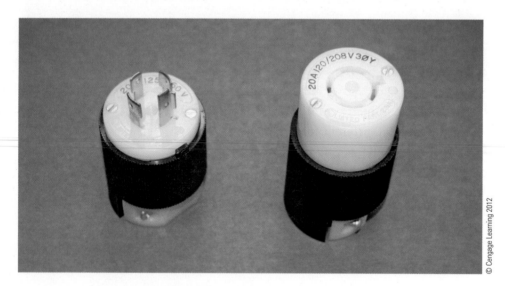

© Cengage Learning 2012

1. Plugs are installed on rubber cord to connect equipment to a receptacle.

2. 90-degree plugs are used where space is limited or when the weight of a straight plug and cord would put excessive pressure on the connection to the receptacle (larger plugs and cords).

3. Twist-lock plugs and connectors are pushed into a receptacle or connector and then twisted clockwise. Likewise, a plug must be twisted counterclockwise to be removed. This function prevents the plug from falling out of the receptacle or connector.

NEC® Article 406, Receptacles, Cord Connectors, and Attachment Plugs

CHAPTER

10

UTILIZATION EQUIPMENT

INTRODUCTION

This chapter covers utilization equipment with its primary focus being on luminaires and their associated equipment. The *NEC®* definition of utilization equipment is *Equipment that utilizes electric energy for electronic, electromechanical, chemical, heating, lighting, or similar purposes. NEC® 100*

The *National Electrical Code* no longer uses the term *lighting fixture* as it has been replaced with the term *luminaire.*

Keep in mind that all equipment must be installed according to the manufacturer's instructions. Manufacturers may have requirements that are over and above *NEC®* requirements that we must follow to properly install the equipment. This is required by the *National Electrical Code®* in *Article 110.3(B).*

This chapter covers some of the luminaires, lamps, and equipment commonly used in the electrical industry. It is by no means complete, as an entire book could be dedicated to the equipment that an electrician will run across; it simply provides an introduction by covering the basics.

Lampholders

Pullchain Lampholder   Keyless Lampholder

© Cengage Learning 2012

1. Lampholders are available in porcelain, plastic, or fiberglass materials.
2. Pullchain lampholders contain a switch that is operated by pulling a chain.
3. Keyless lampholders require an external switch to turn them on and off.
4. Lampholders are available with a built-in receptacle.

NEC® Article 410, Luminaires, Lampholders, and Lamps

Interior Luminaires

Surface-Mount Luminaire

Pendant Luminaire

© Cengage Learning 2012

Recessed Luminaires

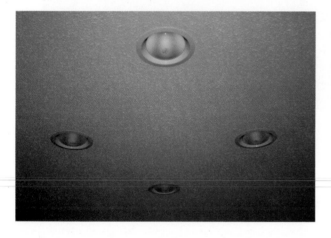

Chain-Hung Luminaire

© Cengage Learning 2012

1. Surface-mount luminaires are available in many styles and types to suit the need of the customer.
   a. These fixtures come with a layer of insulation that must be used to prevent the heat generated in the fixture from overheating the conductor insulation in the box above.
2. Pendant luminaires have a rod or cord that suspends them from the ceiling.
3. Recessed luminaires consist of a can or housing that is above the sheetrock, and a trim that slips up into the housing. Recessed luminaires are available in many sizes and with various trims.
   a. Recessed cans must not be in contact with insulation unless they are "IC rated."
4. The height of chain-hung luminaires can be adjusted by adding or removing lengths of chain. Some chain-hung fixtures are extremely heavy and will need a box listed for the extra weight, or they must be supported independently of the box.

NEC® Article 410, Luminaires, Lampholders, and Lamps

Interior Luminaires (continued)

Track Light

Wall Light

Undercabinet Light

Puck Light

1. Track lighting consists of a track that is mounted to the structure, and lighting heads that can be inserted into the track wherever the light is needed.
2. Track lights are available with multiple circuits in the track so individual light heads can be controlled by separate switches.
3. Undercabinet lighting is a form of task lighting that is typically used to light countertop areas in kitchen and laundry rooms.
   a. Undercabinet lighting is available in several lengths and with various types of lamps.
4. Puck lights are accent lights that are typically installed in cabinets, hutches, alcoves, and so on, to showcase items. They are typically a low-voltage system and require a transformer to step the voltage down.

NEC® Article 410, Luminaires, Lampholders, and Lamps

Exterior Luminaires

Ceiling Light

© Cengage Learning 2012

Wall Light

Post Light

© Cengage Learning 2012

1. Exterior ceiling lights are typically mounted under roof overhangs or in open porches. They must be listed for use in a damp location.
2. Exterior wall lights and post lights must be listed for use in wet locations, as they will be exposed to rain.
3. Post lights will occasionally have a receptacle built into the post, to allow a convenience receptacle in the yard.

NEC® Article 410, Luminaires, Lampholders, and Lamps

Recessed Luminaire Components

Recessed Fixture

Remodeler (Cut-in) Recessed Fixture

Recessed Luminaire Trim

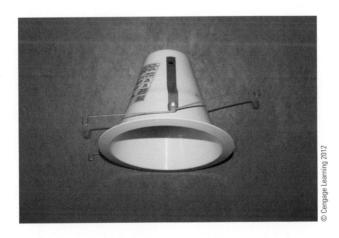

1. Recessed luminaires consist of a recessed fixture, a trim, and a lamp.
   a. Recessed luminaires are available in a variety of styles and sizes.
   b. Recessed luminaires are available with gaskets to create a vapor seal when required by building codes.
   c. Recessed fixtures that have an "IC" rating may be installed in contact with insulation.
2. Remodeler recessed fixtures can be cut into an existing ceiling.
3. Recessed lighting fixtures have a thermal switch in the housing that will remove the power if the enclosure becomes too hot.
4. Recessed trims are available in a variety of styles.

NEC® Article 410, Luminaires, Lampholders, and Lamps

Fluorescent Luminaires

Surface Mount Fluorescent with Reflector Shield

Recessed Fluorescent (Troffer) with Parabolic Lens

Electronic Ballast

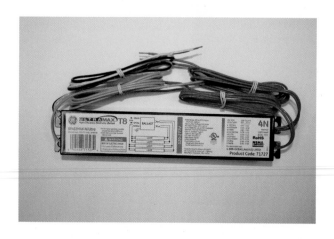

Fluorescent Socket

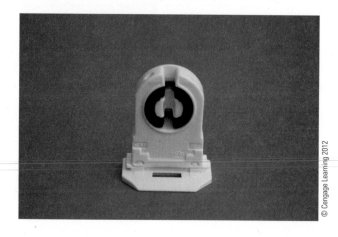

1. Fluorescent lights are available in many styles.
2. Surface-mount fluorescent may have exposed lamps or have a lens that wraps around the fixture that diffuses the light.
3. Fluorescent lights that are designed to be installed into a suspended ceiling are known as "Troffer" fixtures.
4. The typical size for a troffer fixture is 2' × 2' or 2' × 4'.
5. Parabolic lenses are available for fluorescent lights to help prevent reflection and glare on computer screens.
6. Ballasts are the component that drives the fluorescent lamps.
7. Sockets hold fluorescent lamps in place and make their electrical connection.

NEC® Article 410, Luminaires, Lampholders, and Lamps

High-Intensity Discharge (HID) Luminaires

High Bay

Wall Pack

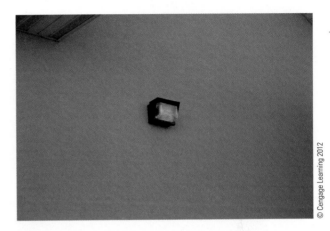

© Cengage Learning 2012

1. HID lighting produces a large amount of light that can be spread over a large area, making them popular for lighting in large buildings and for outdoor lighting.

2. High bay lighting fixtures are used in buildings with high ceilings, where the light has plenty of room to spread out.

3. Wall pack fixtures are mounted on the outside of buildings. They may have a photocell mounted on the fixture to turn it on and off.

NEC® Article 410, Luminaires, Lampholders, and Lamps

Lamp Bases

Candelabra   Medium   Mogul

© Cengage Learning 2012

1. Candelabra base lamps are often found in decorative incandescent fixtures.
2. Medium base are the most commonly used base for incandescent and compact fluorescent luminaires.
3. Mogul base lamps are often found in HID lighting.

NEC® Article 410, Luminaires, Lampholders, and Lamps

Lamps

Incandescent   Compact Fluorescent (CFL)   Light-Emitting Diode (LED)

© Cengage Learning 2012

1. Incandescent lamps have been around for a long time and are still commonly used, but they don't perform as well as some of the other lamps on the market.
    a. Incandescent lamps produce a warm light that is pleasing in homes.
    b. Incandescent lamps are sensitive to vibration.
    c. Incandescent lamps are hot to the touch.
    d. Incandescent lamps are low efficiency.
    e. Incandescent lamps have a short life expectancy (1200 hours).
    f. Incandescent lamps are inexpensive.
2. Compact fluorescent (CFL) lamps are becoming more popular.
    a. CFLs are available in a wide spectrum of colors (2700 K being the closest to incandescent).
    b. CFLs take a few minutes to warm up and produce full intensity.
    c. CFLs are cool to the touch.
    d. CFLS are high efficiency.
    e. CFLs have a long life expectancy (8000 hours).
    f. CFLs have a moderate cost.
3. Light-emitting diode (LED) lamps are fairly new to the general illumination scene. As technology continues to improve, they will become a great option.
    a. LED lamps are available in a wide spectrum of colors (2700 being the closest to incandescent).
    b. LED lamps are cool to the touch.
    c. LED lamps are high efficiency.
    d. LED lamps have the longest life expectancy (50,000 hours).
    e. LED lamps are expensive.

NEC® Article 410, Luminaires, Lampholders, and Lamps

Flood Lamps

R-20   R-30   PAR-30   R-40

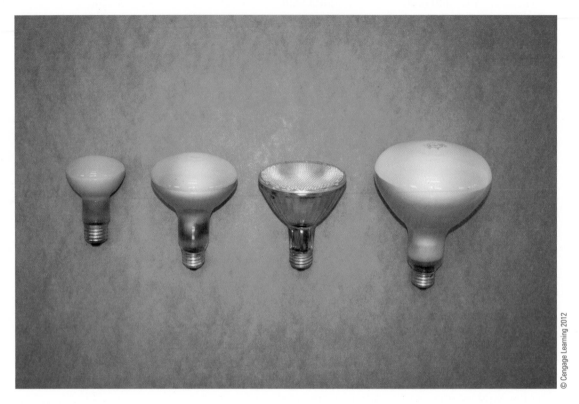

© Cengage Learning 2012

1. Flood lamps will direct the light in the direction the lamp is facing. The "R" stands for reflector, as the interior of the lamp is reflecting and directs the light out the front.

2. The number in flood lamps represents the diameter of the lamp.
    a. 20   2.5" diameter
    b. 30   3.75" diameter
    c. 40   5" diameter

3. Parabolic aluminized reflector (PAR) lamps are often used outside in floodlight-type fixtures.

4. Flood lamps are available in incandescent, halogen, compact fluorescent, and LED.

NEC® Article 410, Luminaires, Lampholders, and Lamps

Fluorescent Lamps

T-5 Lamp   T-8 Lamp   T-12 Lamp

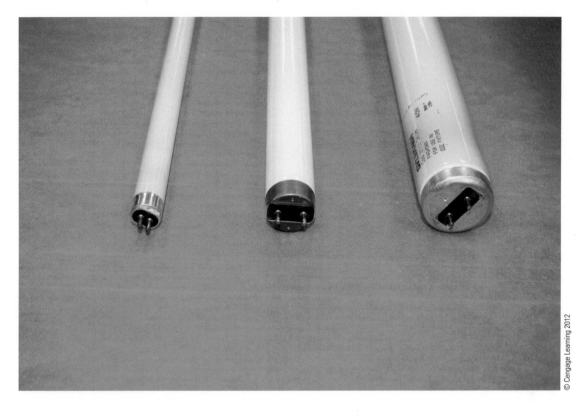

© Cengage Learning 2012

1. Fluorescent lamps come in several diameters and lengths.
2. The number in the name of the lamp represents the lamp diameter in 1/8's of an inch.
   a. T-5   5/8" diameter
   b. T-8   1" diameter
   c. T-12   1½" diameter
3. T-5 and T-8 lamps are used with higher efficiency electronic ballasts.
4. Fluorescent lamps are available in a variety of colors and are represented by their Kelvin (K) number. The lower the number, the yellower the light, and the higher the number, the whiter the light.
   a. 2700K is comparable to the color of incandescent light.
   b. 4100K is a popular color for offices and schools as it is a whiter light and easier on the eyes when reading.
   c. 6500K is compared to daylight on an overcast day.

NEC® Article 410, Luminaires, Lampholders, and Lamps

High-Intensity Discharge (HID) Lamps

High-Pressure Sodium   Mercury Vapor   Metal Halide

© Cengage Learning 2012

1. High-pressure sodium lamps generally produce a yellow light and are often used in exterior lighting.
2. Mercury vapor lamps produce a white light and are often used in outdoor lighting such as yard lights and streetlights.
3. Metal halide lamps produce a white light and are used in high bay fixtures, low bay fixtures, and exterior lighting.

NEC® Article 410, Luminaires, Lampholders, and Lamps

Fans

Paddle Fan

Bathroom Exhaust Fan

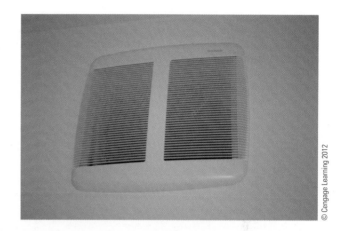

© Cengage Learning 2012

1. Residential paddle fans are occasionally installed in bedrooms and rooms with high ceilings to help move the warm air down from the ceiling in the winter and provide air movement in the summer.

    a. Paddle fans must be installed on a box that is listed for fan support.

    b. UL and manufacturers' instructions require the blades for paddle fans to be at least 7 feet above the floor.

    c. It is recommended to have the fan draw air up in the winter and blow air down in the summer.

    d. Paddle fans typically have a pull chain on the fan to change the speed.

    e. Light kits are available.

2. Bathroom exhaust fans come in many varieties and styles. Combination units are available that may include any combination of the following: fan, light, night light, heat lamps, fan-forced heater.

    a. Exhaust fans are rated in the amount of air in cubic feet per minute (CFM) that they exhaust outside. They will often have a recommendation for the maximum size bathroom that they can efficiently exhaust air from.

    b. Exhaust fans are also rated in sones, the amount of noise the fan makes when it is running. The lower the sone rating, the quieter the fan.

NEC® Article 422, Appliances

Chime Kit

Transformer   Cover

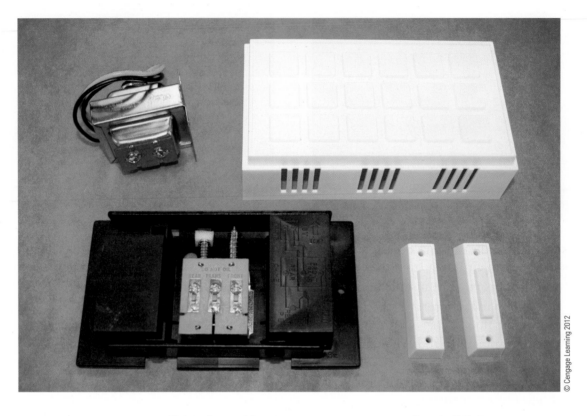

Chime Assembly   Doorbell Buttons

© Cengage Learning 2012

1. Chime kits are typically used in residential applications to indicate when a person is at the door. The kits contain a chime that creates an audible tone, a button that is a momentary contact switch, and a transformer that lowers the voltage available in the electrical panel down to the voltage that operates the chime.

Emergency and Exit Lights

Exit Light

Emergency Light

1. Exit lights are used to indicate the location of room or building exits. Many have battery backup so that in the event of a power outage, the occupants can still see the means of egress.

2. Emergency lights have battery backup and provide illumination in the event of a power outage. The battery in the unit charges under normal power conditions.

CHAPTER

CONTROL EQUIPMENT

INTRODUCTION

Control equipment is the components used to automatically control and/or monitor a process. It may be as simple as turning on a light or as complex as running an assembly line. This chapter covers some of the basic components used in control circuitry.

Time Clock and Photocells

Time Clock

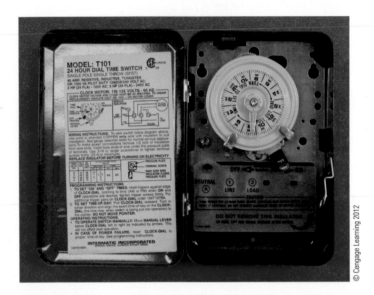

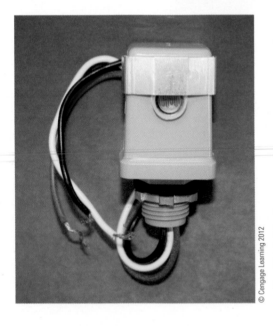

Photocell (Arm Type)

Photocell (Button Type)

1. Time clocks turn loads on and off according to the time of day. They are often used for parking lots or exterior lighting.
2. Photocells turn loads (typically lights) on and off according to the amount of light present.
3. Arm-type photocells can be installed in a threaded hub or in a knockout.
4. Button-type photocells are installed in a knockout (typically in the side of a wall pack exterior lighting fixture).

Thermostats

Low-Voltage Thermostat

Line-Voltage Thermostat

1. Thermostats open or close a set of contacts according to temperature. Some thermostats close when temperatures increase, and some close when temperatures decrease.

2. Thermostats are available in low voltage and line voltage.

3. Low-voltage thermostats typically control central heating equipment such as furnaces and boilers.

4. Line-voltage thermostats typically control electric heating equipment such as baseboard heaters, cove heaters, and so on.

Push Buttons

Push/Pull Operator   Flat Push Button   Extended Push Button

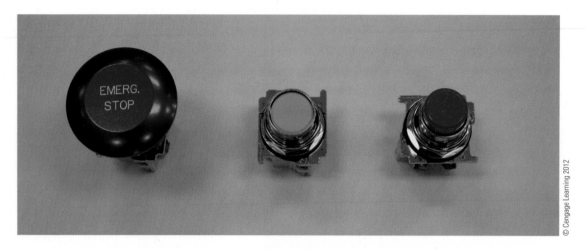

Push-Button Stations

1. Push/pull operators have maintained contacts and often have a mushroom head.
2. Push buttons typically have momentary contact.
3. Push buttons are available with normally open or normally closed contacts.
4. Push buttons are available in several colors and may be illuminated.
5. Push-button stations have push buttons built into an enclosure. They are available with several faces on the pushbuttons: (start, stop) (open, close, stop) (forward, reverse, stop).

Selector Switches and Indicator Lights

Selector Switches

Indicator Lights

1. Selector switches are available with normally open or normally closed contacts.
2. Selector switches may have momentary contacts or maintained contacts.
3. Selector switches may have two, three, or four positions.
4. Indicator lights are used to indicate when a load or process is on.
5. Indicator lights are available in many colors.

Limit Switches and Sensors

Limit Switch with Roller Operator

Photo Switch

Proximity Sensor

1. Limit switches, photo switches, and proximity sensors indicate the position of a product.
2. Limit switches are available with normally open or normally closed contacts.
3. Limit switches are available with multiple numbers of contacts.
4. Limit switches are available with several types of operators.

Relays/Contactors

RIB (Relay in a Box®)

Two-Pole Relay with N/O and N/C Contacts

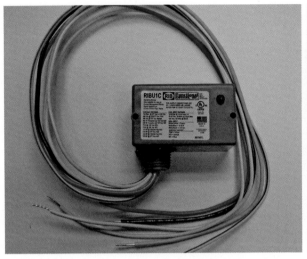

© Cengage Learning 2012

1. Relays are used to turn loads on or off.
2. Larger relays are often called *contactors*.
3. Relays called Relay in a Box® (RIB) already have a box, so they do not need to be installed in a separate box. The disadvantage of using a RIB, however, is that its contact ampacity is relatively small.
4. Relays are available in a 2-pole version that has two sets of contacts that are controlled together.
5. Relays are available with many coil voltages and contact configurations:
   a. 1-pole, 2-pole, 3-pole, and so on.
   b. normally open contacts, normally closed contacts, or both.
6. Relays and contactors are available in ampacities from less than an amp to several hundred amps.

Relays/Contactors (continued)

Base   Plug-in Relay (Cube Relay)   Programmable Relay (Smart Relay)

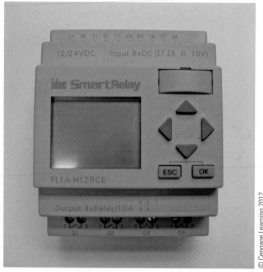

© Cengage Learning 2012

1. Cube relays are easy to replace: Simply pull the old relay out of the base and plug in a new one; no connections have to be removed.

2. Programmable relays can be used as timers, relays, time clocks, and much more.

3. Programmable relays can be programmed to perform functions related to time or inputs. For example, inputs can be pushbuttons, limit switches, photocells, and so on.

4. Programmable relays are similar to programmable logic controllers.

Motor Starters

NEMA Motor Starter

IEC Motor Starter

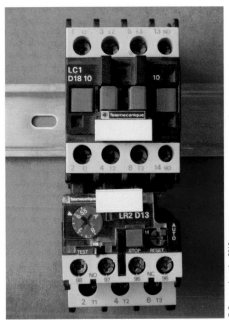

© Cengage Learning 2012

1. Motor starters are used to turn motors on and off and to provide overload protection.
    a. As power is applied to the coil of the motor starter, it closes a set of contacts, allowing current to pass to the motor.
    b. Motor starters have overload protection (sometimes called *heaters*). If excess current runs to the motor, the overload protection opens the motor circuit.
2. Motor starters vary slightly in size: IEC motor starters are generally smaller than the equivalently rated NEMA motor starters and often are designed to be attached to a din rail.

Programmable Logic Controller

Modular PLC

Fixed PLC

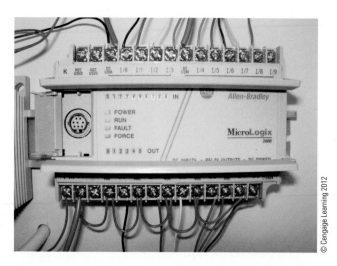

1. Programmable logic controllers (PLCs) are used to control processes such as those in industrial plants or assembly lines. They contain computer processors that run a program determining how a process works.

2. PLCs have inputs and outputs:
   a. Inputs indicate status or give a signal to the PLC and include:
      i. start pushbuttons
      ii. stop pushbuttons
      iii. flow switches
      iv. proximity switches
   b. Outputs control the process and provide information to the operator and include:
      i. indicator lights
      ii. motor starters
      iii. solenoids

3. PLCs allow for added inputs and outputs in the modular version.

4. PLCs have a fixed number of inputs and outputs in the fixed version.

Off-Peak/Ripple Equipment

Load Controller (Ripple Receiver)

Current Transformers

1. Load controllers receive a signal from a utility company to turn customer loads on or off according to utility demand. The signal used to communicate with the ripple receiver may be radio frequency or a frequency in the power line.

2. Current transformers (CT) measure the amount of current passing through a conductor. The type of CT shown above typically is used with an off-peak system to measure the amount of power used to feed loads for a furnace, a water heater, an AC unit, a boiler, and so on.

CHAPTER

COMMUNICATIONS

INTRODUCTION

Communication equipment and wiring is a continually changing part of the electrical industry. Items that were brand new this year will be obsolete next year. This chapter will introduce some of the items that electricians will run into that have been commonly used throughout the years.

Punch-Down Blocks (Type 66 Blocks)

Punch-Down Blocks (Type 66 Blocks)

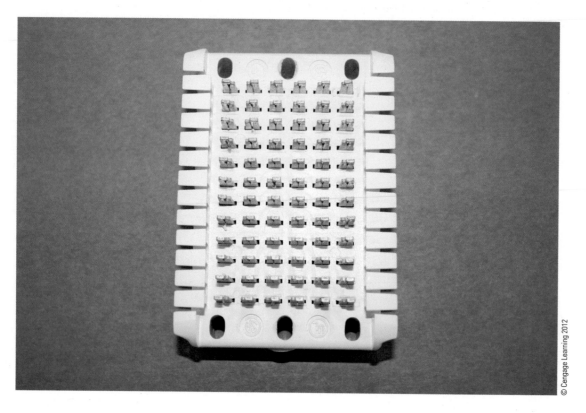

© Cengage Learning 2012

1. Type 66 terminal blocks are typically used to terminate voice circuits. As the conductor is pushed into the block's metal tabs, the insulation is displaced to make contact between the conductor and the terminal.

Communication Jacks
Wall-Hanging Jacks

Telephone Plate   TV Plate   Combination Plate   Telephone Plate

© Cengage Learning 2012

1. Telephone plates are available with single or dual jacks.
2. Combination plates contain a telephone jack and a TV jack.
3. TV plates are available with single or dual jacks.
4. Wall-hanging telephone plates have studs on which to hang a telephone.
5. TV and telephone plates are available in a variety of colors.

Communication Jacks (continued)
Surface-Mounted Telephone Jacks

© Cengage Learning 2012

1. Surface-mount data and telephone jacks are often used to add a line in a finished area of a building.
2. Using a surface-mount jack makes it possible to avoid the hassle of having to cut a hole in the finished wall surface and fish a wire in the wall.

Modular Plates and Ports

Four-Position Plate   Coaxial Modular Port

Cat 5 Modular Port

1. Modular plates contain openings for modular ports to snap into, which enables any combination of jacks to be available in a plate.

2. Modular plates are available in many configurations, containing from one to eight openings.

3. Modular ports are available for many types of jacks, including:
   a. coaxial
   b. Cat 3
   c. Cat 5
   d. RCA

TV Connectors and Splitters

F-Connector   Eight-Way Splitter   Two-Way Splitter

© Cengage Learning 2012

1. F-connectors are crimped onto the ends of coaxial cable to terminate the cable at a jack or a splitter.
2. Splitters split a TV signal to multiple locations.
3. Splitters are available with many numbers of outputs.

Communications Straps and Fasteners

Cable Straps

© Cengage Learning 2012

Bridle Ring   J-Hook

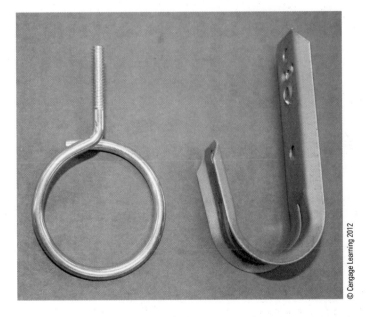

© Cengage Learning 2012

Slotted Wall-Wiring Duct

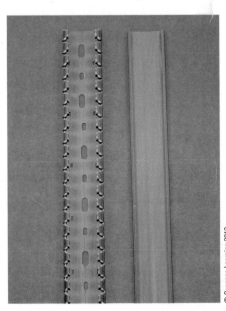

© Cengage Learning 2012

1. Cable straps bundle or tie communication cables without damaging the cable conductors.
2. Bridle rings and J-hooks support communication cables.
3. Slotted wall-wiring duct is used as a surface raceway for communication cables. The duct's cover is removable, providing easy access to the raceway to add or remove cables.